Texts in Mathematics

Volume 6

A Mathematical Primer on Linear Optimization

Texts in Mathematics Series Editor
Dov Gabbay
                                            dov.gabbay@kcl.ac.uk

A Mathematical Primer on Linear Optimization

Diogo Gomes
Amílcar Sernadas
Cristina Sernadas
João Rasga
Paulo Mateus

ISBN 978-1-84890-315-9

College Publications
Scientific Director: Dov Gabbay
Managing Director: Jane Spurr
Department of Computer Science
King's College London, Strand, London WC2R 2LS, UK

http://www.collegepublications.co.uk

Cover designed by Laraine Welch

Preface

Our main objective is to provide a self-contained mathematical introduction to linear optimization for undergraduate students of Mathematics. This book is equally suitable for Science, Engineering, and Economics students who are interested in gaining a deeper understanding of the mathematical aspects of the subject. The linear optimization problem is analyzed from different perspectives: topological, algebraic, geometrical, logical, and algorithmic. Nevertheless, no previous knowledge of these subjects is required. The essential details are always provided in a special section at the end of each chapter. The technical material is illustrated with multiple examples, problems with fully-worked solutions, and a range of proposed exercises.

In Chapter 1, the optimization problem is presented along with the concepts of set of admissible vectors and set of optimizers. Then, we discuss the linear case, including the canonical and the standard optimization problems. Finally, we relate the general linear optimization problem and the canonical optimization problem and analyze the relationship between the canonical and the standard optimization problems. The relevant background section of this chapter includes some basic algebraic concepts, notation, and preliminary results namely about groups, fields, and vector spaces.

Chapter 2 explores some topological techniques that provide sufficient conditions for the existence of optimizers in canonical optimization problems. The chapter starts with the definition of interior and boundary of the set of admissible vectors (later on it is shown that these concepts coincide with the topological ones) and the proof that optimizers are always on the boundary of the set of admissible vectors when the objective map is not the zero map. Then, we provide sufficient conditions for a canonical optimization problem to have maximizers. In the relevant background section, we provide a *modicum* of topological notions and results.

The main objective of Chapter 3 is to provide a way for deciding whether or not an admissible vector is an optimizer, relying on Farkas' Lemma. Namely, we introduce the concept of line active in an admissible vector and prove the Local Maximizer Theorem and the Maximizer Theorem using convex cones. Moreover, a technique for deciding whether or not there are admissible vectors is presented, using Farkas' Lemma again.

In Chapter 4, linear algebra concepts are used for computing optimizers for a standard optimization problem. The objective is to find the basic admissible vectors of the problem at hand. These vectors are relevant since we prove that, under certain conditions, there is always a basic admissible vector which is an optimizer. Moreover, the notion of a non-degenerate standard optimization problem is introduced and a characterization of basic vectors is provided in this case. In the relevant background section of this chapter, we provide an overview of notions and results from linear algebra including dimension, span, rank, determinant of a matrix, singular matrices, and some well-known theorems.

Chapter 5 concentrates on geometrical aspects of linear optimization, namely it provides a geometrical characterization of the basic admissible vectors of a standard optimization problem as vertices of an appropriate convex polyhedron. In the relevant background section, we present the relevant results about affine spaces and subspaces.

Chapter 6 presents several aspects of duality. The objective is to show that the dual provides yet a new technique for finding optimizers of the original problem. The dual problems are found using the Lagrange Multiplier Technique. We start by presenting two main results: the Weak and the Strong Duality Theorems. Then, we concentrate on slacks for the linear optimization problem and prove the Slack Complementarity Theorem. Afterward, the Equilibrium Theorem is established as well as the Uniqueness Theorem. Finally, a Logic of Inequalities is presented and some results are proved relating consistency and satisfiability of a formula.

Chapter 7 provides an introduction to computational complexity to analyze the efficiency of linear optimization algorithms. In Section 7.1, we present the rigorous definition of the decision problems associated with linear optimization, and in Section 7.2, we discuss the representation of vectors and matrices. In Section 7.3, we prove that the standard decision problem is in $\mathcal{NP}$. A deterministic algorithm for getting an optimizer whenever there is one, based on a *brute-force* method, is discussed in Section 7.4. The complexity of this algorithm is shown not to be polynomial. Finally, in the relevant background section, we introduce the central notions and techniques that are necessary for assessing the computational efficiency of an algorithm.

Chapter 8 is targeted at the Simplex Algorithm that allows us to find an optimizer, whenever there is one, for the standard optimization problem satisfying some mild conditions. In Section 8.2, we prove the soundness and completeness of the Simplex Algorithm and point out that its complexity is not polynomial.

Finally, in Chapter 9, a description of the integer linear optimization problem is presented along with the respective relaxed problem. Then, we compare the sets of admissible vectors and optimizers of both problems. The integrality gap is introduced and discussed. Afterward, we discuss totally unimodular problems and provide a sufficient condition for a matrix to be totally unimodular. Furthermore, we give the Assignment Problem as an example. The importance of this concept is made clear by proving a sufficient condition for the existence and characterization of the optimizers of totally unimodular problems. We conclude by presenting and illustrating an algorithm based on the Branch and Bound Technique for solving integer optimization problems.

Acknowledgements

We would like to express our deepest gratitude to the many undergraduate math students of Instituto Superior Técnico that attended the Introduction to Optimization course.

Diogo Gomes was partially supported by King Abdullah University of Science and Technology (KAUST) baseline funds and KAUST OSR-CRG2017-3452. Amílcar Sernadas, Cristina Sernadas and João Rasga acknowledge the National Funding from Fundação para a Ciência e a Tecnologia (FCT) under the project UID/MAT/04561/2019 granted to Centro de Matemática, Aplicações Fundamentais e Investigação Operacional (CMAFcIO) of Universidade de Lisboa. Paulo Mateus acknowledges the National Funding from FCT under project PEst-OE/EEI/LA0008/2019 granted to Instituto de Telecomunicações.

Last but not least, we greatly acknowledge the excellent working environment provided by the Department of Mathematics of Instituto Superior Técnico, Universidade de Lisboa.

Lisbon,
July 2019

Diogo Gomes
Amílcar Sernadas
Cristina Sernadas
João Rasga
Paulo Mateus

Contents

Chapter 1

Optimization Problems

The objective of this chapter is to present the optimization problem and its different formulations.

1.1   General Concepts

In this section, we introduce the general form of an optimization problem. Furthermore, we discuss admissible tuples, maximizers and minimizers.

Definition 1.1
An *(n-dimensional) constraint*, with $n \in \mathbb{N}^+$, is a triple $(g, \bowtie, b)$, written

$$g(x_1, \ldots, x_n) \bowtie b,$$

where $g : \mathbb{R}^n \to \mathbb{R}$ is a *constraint map*, $\bowtie \in \{\leq, =, \geq\}$ is a binary relation and $b \in \mathbb{R}$. When $\bowtie$ is $=$, we say that the constraint is an *equality* and when $\bowtie$ is either $\leq$ or $\geq$, we say that the constraint is an *inequality*.

Definition 1.2
An *(n-dimensional) description*, with $n \in \mathbb{N}^+$, is a pair

$$(\{(g_i, \bowtie_i, b_i) : 1 \leq i \leq m\}, U)$$

for some $m \in \mathbb{N}^+$, where $(g_i, \bowtie_i, b_i)$ is an n-dimensional constraint for $i = 1, \ldots, m$ and $U \subseteq \mathbb{R}$ is a non-empty set.

When $U = \mathbb{R}$ in an n-dimensional description, we may simply present the description by the set of constraints.

1

Example 1.1 (Rectangle Problem)
Assume that we want to consider all rectangles with a fixed area $b \in \mathbb{R}^+$. This universe can be expressed as a 2-dimensional description

$$\{x_1 x_2 = b, x_1 \geq 0, x_2 \geq 0\},$$

where x_1 and x_2 represent the dimensions (length and width) of the rectangles.

Example 1.2 (Knapsack Problem)
Assume that we have n kinds of objects that we want to accommodate in a knapsack that has a weight limit b. These requirements can be stated as an n-dimensional description

$$(\{a_1 x_1 + \cdots + a_n x_n \leq b\}, \mathbb{N}),$$

where each x_j represents the number of objects of kind j and a_j is the weight of each object of kind j.

Definition 1.3
We say that $(d_1, \ldots, d_n) \in U^n$ is an *admissible vector* for an n-dimensional description

$$(\{g_i(x_1, \ldots, x_n) \bowtie b_i : 1 \leq i \leq m\}, U)$$

when

$$g_i(d_1, \ldots, d_n) \bowtie b_i$$

holds for every $i = 1, \ldots, m$.

Notation 1.1
We denote by

$$X_D$$

the set of admissible tuples or vectors for a description D.

Example 1.3 (Rectangle Problem)
Let D be the description introduced in Example 1.1. Then,

$$\left(\sqrt{b}, \sqrt{b}\right) \in X_D \quad \text{and} \quad \left(\frac{b}{7}, \frac{b}{7}\right) \notin X_D.$$

Definition 1.4
An *(n-dimensional) optimization problem* P is a tuple

$$(D, f, \diamond),$$

where D is an n-dimensional description, $f : \mathbb{R}^n \to \mathbb{R}$ is the *objective map* and $\diamond \in \{\leq, \geq\}$. We denote by

$$S_P$$

the set of *optimizers* for P composed by each $(s_1, \ldots, s_n) \in X_D$ such that

$$f(d_1, \ldots, d_n) \diamond f(s_1, \ldots, s_n),$$

for every $(d_1, \ldots, d_n) \in X_D$.

Notation 1.2
Given an optimization problem $(D, f, \diamond)$, we denote by

$$X_P$$

the set X_D. Moreover, when $\diamond$ is $\leq$, P is a *maximization problem* and each element of S_P is a *maximizer* of P. Otherwise, P is a *minimization problem* and each element of S_P is a *minimizer* of P.

Example 1.4 (Rectangle Problem)
Recall the 2-dimensional description D introduced in Example 1.1. Let $f : \mathbb{R}^2 \to \mathbb{R}$ be such that

$$f(x_1, x_2) = 2x_1 + 2x_2.$$

Assume that we want to find a rectangle with the minimum perimeter among the rectangles satisfying the description D. This can be expressed by the minimization problem:

$$P = (D, f, \geq).$$

We now show that the rectangle $(\sqrt{b}, \sqrt{b})$ is a minimizer of P. Observe that if $(d_1, d_2) \in X_P$ then $d_1 = \frac{b}{d_2}$; that is,

$$X_P = \left\{ \left(\frac{b}{d}, d \right) : d \in \mathbb{R}^+ \right\}.$$

Let $h : \mathbb{R}^+ \to \mathbb{R}$ be such that

$$h(d) = f\left(\frac{b}{d}, d \right) = 2\frac{b}{d} + 2d.$$

Then, finding a minimizer of P is the same as finding a value for which h is minimal. Taking the derivative h' of h we have:

$$h'(d) = -2\frac{b}{d^2} + 2$$

which has a unique zero in $\mathbb{R}^+$ at $d = \sqrt{b}$. Taking into account that the second derivative of h at d is positive then $\sqrt{b}$ is a minimizer of P (see [3]).

Remark 1.1

In the sequel, we may use

$$x \mapsto f(x)$$

for presenting a map f.

Example 1.5 (Knapsack Problem)

Recall the n-dimensional description D introduced in Example 1.2. Assume that each object of kind j has value c_j. The *knapsack problem* consists in maximizing the value of objects that can be put in the knapsack respecting the limit weight. Hence, it can be described as the optimization problem

$$(D, f, \le),$$

where $f : (x_1, \ldots, x_n) \mapsto c_1 x_1 + \cdots + c_n x_n$. In Chapter 9, we shall return to this problem.

Notation 1.3

In the sequel, we present the n-dimensional optimization problem

$$(((\{g_i(x_1, \ldots, x_n) \bowtie b_i : 1 \le i \le m\}, U), f, \le)$$

by

$$\begin{cases} \max\limits_{(x_1, \ldots, x_n)} f(x_1, \ldots .x_n) \\ g_1(x_1, \ldots .x_n) \bowtie_1 b_1 \\ \vdots \\ g_m(x_1, \ldots .x_n) \bowtie_m b_m \\ x_1, \ldots, x_n \in U \end{cases}$$

or even

$$\begin{cases} \max\limits_{x} f(x) \\ g_1(x) \bowtie_1 b_1 \\ \vdots \\ g_m(x) \bowtie_m b_m \\ x \in U^n. \end{cases}$$

Similarly for $\ge$ using min instead of max. When $U = \mathbb{R}$, we may omit the constraint on U.

Example 1.6 (Knapsack Problem)
The n-dimensional optimization problem introduced in Example 1.5 can be presented as follows:

$$
\begin{cases}
\displaystyle\max_{(x_1,\ldots,x_n)} \; c_1 x_1 + \cdots + c_n x_n \\
a_1 x_1 + \cdots + a_n x_n \leq b \\
x_1, \ldots, x_n \in \mathbb{N}.
\end{cases}
$$

We can present the optimization problem in a more condensed way. For that, we need to introduce the following relations.

Definition 1.5
Let $\leq$ be the binary relation over $\mathbb{R}^n$ such that

$$x \leq y$$

whenever $x_j \leq y_j$ for every $j = 1, \ldots, n$. Similarly, for $\geq$ and $=$.

Example 1.7
As an illustration, observe that $(0,1) \leq (2,3)$. Nevertheless, $\leq$ is not a total relation. For example

$$(-3,5) \not\leq (1,0) \quad \text{and} \quad (1,0) \not\leq (-3,5);$$

that is, $(-3,5)$ and $(1,0)$ are not comparable.

Exercise 1.1

Present x and y in $\mathbb{R}^n$ such that

- $x \neq y$;

- $x \not\leq y$;

- $x \not\geq y$.

Notation 1.4
Sometimes, it is convenient to group constraints by their associated relational

symbols; that is, given an n-dimensional optimization problem of the form

$$\begin{cases} \max\limits_{(x_1,\ldots,x_n)} \quad f(x_1,\ldots,x_n) \\ g_i'(x_1,\ldots,x_n) \leq b_i', \quad \text{for } i = 1,\ldots,m' \\ g_i''(x_1,\ldots,x_n) \geq b_i'', \quad \text{for } i = 1,\ldots,m'' \\ g_i'''(x_1,\ldots,x_n) = b_i''', \quad \text{for } i = 1,\ldots,m''' \\ x_1,\ldots,x_n \in U, \end{cases}$$

where $m', m'', m''' \in \mathbb{N}$, we denote by

$$g_\leq : \mathbb{R}^n \to \mathbb{R}^{m'}$$

the $\leq$-*constraint map* such that

$$g_\leq(x_1,\ldots,x_n) = (g_1'(x_1,\ldots,x_n),\ldots,g_{m'}'(x_1,\ldots,x_n)).$$

Similarly, for $g_\geq$ and $g_=$. Thus, the previous problem can be presented as follows:

$$\begin{cases} \max\limits_{(x_1,\ldots,x_n)} \quad f(x_1,\ldots,x_n) \\ g_\leq(x_1,\ldots,x_n) \leq b' \\ g_\geq(x_1,\ldots,x_n) \geq b'' \\ g_=(x_1,\ldots,x_n) = b''' \\ x_1,\ldots,x_n \in U, \end{cases}$$

where $b' = (b_1',\ldots,b_{m'}')$, $b'' = (b_1'',\ldots,b_{m''}'')$ and $b''' = (b_1''',\ldots,b_{m'''}''')$. Similarly for minimization problems.

1.2　Linear Optimization

There are several versions of optimization problems depending on the properties of the objective map and of the constraint maps (see [55, 10, 29, 32, 11, 53]). In this book, we concentrate on the linear case. The reader can recall the relevant (linear) algebraic notions needed in Section 1.6. We consider the vector spaces $\mathbb{R}^j$ over $\mathbb{R}$ for $j \in \mathbb{N}^+$. Moreover, we assume that the constraint maps and the objective map are linear maps between appropriate vector spaces as we discuss now.

Definition 1.6
We say that a maximization problem

$$\begin{cases} \max_{(x_1,\ldots,x_n)} f(x_1,\ldots,x_n) \\ g_{\le}(x_1,\ldots,x_n) \le b' \\ g_{\ge}(x_1,\ldots,x_n) \ge b'' \\ g_{=}(x_1,\ldots,x_n) = b''' \\ x_1,\ldots,x_n \in U \end{cases}$$

is *general linear* when $f, g_{\le}, g_{\ge}$ and $g_{=}$ are linear maps. Moreover, a general linear problem is *linear* when $U = \mathbb{R}$. Similarly, for minimization problems.

Example 1.8
The optimization problem

$$\begin{cases} \max_{(x_1,x_2)} 2x_1 + x_2 \\ 3x_1 - x_2 \le 6 \\ (x_1 - 3x_2, x_1, x_2) \ge (-6,0,0) \end{cases}$$

is linear.

Notation 1.5
We denote by

$$\mathcal{L}$$

the set of all linear optimization problems.

Remark 1.2
From now on, we present the linear maps in an optimization problem by the induced matrices with respect to the standard basis (see Section 1.6). Moreover, we assume that each line of the induced matrices is non-null.

Notation 1.6
Taking into account the previous remark, any general linear maximization prob-

lem can be presented in matricial form either in an expanded way as:

$$\begin{cases} \max_{x} \; cx \\ A'x \leq b' \\ A''x \geq b'' \\ A'''x = b''' \\ x \in U^n \end{cases}$$

or, in a more compact way, as

$$(A', A'', A''', b', b'', b''', c, \leq, U),$$

where

- $A' = (a'_{ij}) \in \mathbb{R}^{m' \times n}$, $A'' = (a''_{ij}) \in \mathbb{R}^{m'' \times n}$ and $A''' = (a'''_{ij}) \in \mathbb{R}^{m''' \times n}$;

- $b' = (b'_i) \in \mathbb{R}^{m' \times 1}$, $b'' = (b''_i) \in \mathbb{R}^{m'' \times 1}$ and $b''' = (b'''_i) \in \mathbb{R}^{m''' \times 1}$;

- $c = (c_j) \in \mathbb{R}^{1 \times n}$.

Similarly for minimization problems. When $m' = 0$, then A' is the empty matrix and similarly for m'' and m'''. For simplification, when there is no ambiguity, we omit empty matrices and vectors in the expanded presentation. Furthermore, as before, when $U = \mathbb{R}$, we may omit the constraint on U.

Example 1.9
The linear optimization problem presented in Example 1.8 can be described in expanded matricial form as follows:

$$\begin{cases} \max_{x} \begin{bmatrix} 2 & 1 \end{bmatrix} x \\ \\ \begin{bmatrix} 3 & -1 \end{bmatrix} x \; \leq \; \begin{bmatrix} 6 \end{bmatrix} \\ \\ \begin{bmatrix} 1 & -3 \\ 1 & 0 \\ 0 & 1 \end{bmatrix} x \; \geq \; \begin{bmatrix} -6 \\ 0 \\ 0 \end{bmatrix}. \end{cases}$$

In a more compact way, we can write

$$(\begin{bmatrix} 3 & -1 \end{bmatrix}, \begin{bmatrix} 1 & -3 \\ 1 & 0 \\ 0 & 1 \end{bmatrix}, [], \begin{bmatrix} 6 \end{bmatrix}, \begin{bmatrix} -6 \\ 0 \\ 0 \end{bmatrix}, [], \begin{bmatrix} 2 & 1 \end{bmatrix}, \leq).$$

Example 1.10 (Diet Problem)

A dog food production factory wants to minimize the fat included in each meal while complying with specific nutritional requirements establishing that each meal should have at least a certain quantity of essential nutrients. To do so, the nutritionist has decided that each meal should use given food components. Each food component has a certain amount of fat and a certain quantity of nutrients. To model this problem, assume that:

- the important nutrients range from nutrient 1 to nutrient m;

- the least recommended quantity of nutrient i in each meal is b_i;

- the food components that can be included in a meal range from component 1 to component n;

- each food component j contains the quantity a_{ij} of nutrient i;

- each food component j contains c_j unities of fat.

The goal is to minimize the amount of fat in each meal. The only way to do so is by adjusting the amount of each food component in a meal. Thus, let

$$x_j$$

be the amount of food component j in the meal. Hence, the total amount of fat in a meal is given by:

$$\sum_{j=1}^{n} c_j x_j.$$

The goal of the problem is to minimize this quantity taking account the requirements on the nutrients, which are modeled by

$$\sum_{j=1}^{n} a_{ij} x_j \geq b_i$$

for $i = 1, \ldots, m$. Therefore, the linear optimization problem can be presented as follows:

$$\begin{cases} \min_{(x_1,\ldots,x_n)} \sum_{j=1}^{n} c_j x_j \\ \sum_{j=1}^{n} a_{ij} x_j \geq b_i, \quad \text{for } i = 1, \ldots, m \\ x_1, \ldots, x_n \geq 0 \end{cases}$$

or, in matricial form, as follows:

$$
\begin{cases}
\min\limits_{x} \begin{bmatrix} c_1 & \cdots & c_n \end{bmatrix} x \\
\\
\begin{bmatrix} a_{11} & \cdots & a_{1n} \\ \vdots & \ddots & \vdots \\ a_{m1} & \cdots & a_{mn} \\ 1 & \cdots & 0 \\ \vdots & \ddots & \vdots \\ 0 & \cdots & 1 \end{bmatrix} x \geq \begin{bmatrix} b_1 \\ \vdots \\ b_m \\ 0 \\ \vdots \\ 0 \end{bmatrix}.
\end{cases}
$$

Exercise 1.2

Let $A \in \mathbb{R}^{m \times n}$, $b \in \mathbb{R}^m$ and $f : \mathbb{R}^{m+n} \to \mathbb{R}$ be such that

$$f(x_1, \ldots, x_n, z_1, \ldots, z_m) = z_1 + \cdots + z_m.$$

Suppose that we want to find the minimizers

$$(x_1, \ldots, x_n, z_1, \ldots, z_m) \geq 0$$

of f fulfilling $A(x_1, \ldots, x_n) - (z_1, \ldots, z_m) \leq b$. Present a linear optimization formulation of the problem.

Definition 1.7

A linear maximization problem

$$
\begin{cases}
\max\limits_{x} cx \\
A'x \leq b' \\
A''x \geq b'' \\
A'''x = b''' \\
x \in U^n
\end{cases}
$$

is *integer linear* when

- $A' = (a'_{ij}) \in \mathbb{Q}^{m' \times n}$, $A'' = (a''_{ij}) \in \mathbb{Q}^{m'' \times n}$ and $A''' = (a'''_{ij}) \in \mathbb{Q}^{m''' \times n}$;

- $b' = (b'_i) \in \mathbb{Q}^{m' \times 1}$, $b'' = (b''_i) \in \mathbb{Q}^{m'' \times 1}$ and $b''' = (b'''_i) \in \mathbb{Q}^{m''' \times 1}$;

- $c = (c_j) \in \mathbb{Q}^{1 \times n}$;

and $U = \mathbb{N}$. Similarly for minimization problems.

Observe that, in the literature, the constraint

$$x \in \mathbb{N}^n$$

may be replaced by the constraints $x \geq 0$ and $x \in \mathbb{Z}^n$.

Example 1.11 (Knapsack Problem)
The optimization problem introduced in Example 1.6 is integer linear.

1.3 Canonical and standard optimization problems

There are two very important classes of linear optimization problems. In the sequel, we use I to denote the identity matrix.

Definition 1.8
A linear optimization problem is *canonical* if it has the form:

$$\begin{cases} \max_{x} cx \\ Ax \leq b \\ x \geq 0, \end{cases}$$

or, in a compact way, $(A, I, [\,], b, 0, [\,], c, \leq)$.

Notation 1.7
When a problem is known to be canonical, it can simply be presented as a triple of the form:

$$(A, b, c).$$

Notation 1.8
Sometimes, we need to present a canonical optimization problem

$$\begin{cases} \max_{x} cx \\ Ax \leq b \\ x \geq 0 \end{cases}$$

in the following (*pure*) form:

$$\begin{cases} \max_{x} cx \\ \begin{bmatrix} A \\ -I \end{bmatrix} x \le \begin{bmatrix} b \\ 0 \end{bmatrix}. \end{cases}$$

In the sequel, we may write

$$\begin{bmatrix} A \\ -I \end{bmatrix} \quad \text{and} \quad \begin{bmatrix} b \\ 0 \end{bmatrix}$$

as $\underline{A}$ and $\underline{b}$, respectively.

Example 1.12
The following linear optimization problem

$$\begin{cases} \max_{x} 2x_1 + x_2 \\ 3x_1 - x_2 \le 6 \\ -x_1 + 3x_2 \le 6 \\ x \ge 0; \end{cases}$$

that is,

$$\begin{cases} \max_{x} \begin{bmatrix} 2 & 1 \end{bmatrix} x \\ \begin{bmatrix} 3 & -1 \\ -1 & 3 \end{bmatrix} x \le \begin{bmatrix} 6 \\ 6 \end{bmatrix} \\ x \ge 0, \end{cases}$$

is canonical. Its pure form is:

$$\begin{cases} \max_{x} \begin{bmatrix} 2 & 1 \end{bmatrix} x \\ \begin{bmatrix} 3 & -1 \\ -1 & 3 \\ -1 & 0 \\ 0 & -1 \end{bmatrix} x \le \begin{bmatrix} 6 \\ 6 \\ 0 \\ 0 \end{bmatrix}. \end{cases}$$

Exercise 1.3

Let P be a canonical optimization problem. Show that

$$S_P = \{s \in X_P : cs = \sup_{x \in X_P} cx\}.$$

Notation 1.9
We denote by

$$\mathcal{C}$$

the set of all canonical linear optimization problems.

Definition 1.9
A linear n-dimensional optimization problem is *standard* if it has the form

$$\begin{cases} \min_x cx \\ Ax = b \\ x \geq 0; \end{cases}$$

that is, is a tuple of the form $([\,], I, A, [\,], 0, b, c, \geq)$.

Notation 1.10
When a problem is known to be standard, it can simply be presented as a triple of the form:

$$(A, b, c).$$

Example 1.13
The following linear optimization problem

$$\begin{cases} \min_x -2x_1 - x_2 \\ 3x_1 - x_2 + x_3 = 6 \\ -x_1 + 3x_2 + x_4 = 6 \\ x \geq 0; \end{cases}$$

that is,

$$\begin{cases} \min_x \begin{bmatrix} -2 & -1 & 0 & 0 \end{bmatrix} x \\[2mm] \begin{bmatrix} 3 & -1 & 1 & 0 \\ -1 & 3 & 0 & 1 \end{bmatrix} x = \begin{bmatrix} 6 \\ 6 \end{bmatrix} \\[2mm] x \geq 0, \end{cases}$$

is standard.

Notation 1.11
We denote by

$$\mathcal{S}$$

the set of all standard linear optimization problems.

One of the objectives of this book is to provide techniques for establishing the existence and for computing optimizers of linear optimization problems. However, it is essential to get intuition in simple cases by using elementary geometric arguments.

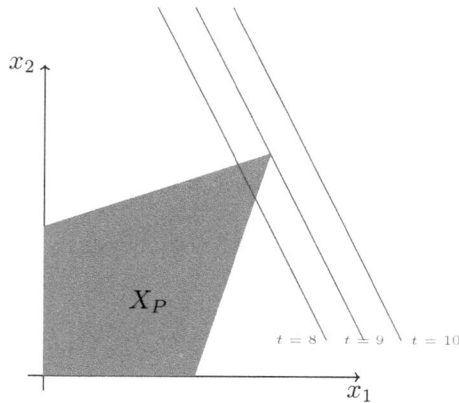

Figure 1.1: Graphical presentation of problem in Example 1.12.

Example 1.14
Denote the canonical optimization problem in Example 1.12 by P. The set X_P

of admissible vectors is depicted in Figure 1.2. We now provide an intuition for concluding that $S_P = \{(3,3)\}$. Since the goal is to maximize $f(x_1, x_2) = 2x_1 + x_2$, we need to find the maximum $t \in \mathbb{R}$ such that the intersection of the line

$$2x_1 + x_2 = t$$

with X_P is non-empty. As can be seen in Figure 1.1, this is achieved for $t = 9$ at $(3,3)$.

1.4   Relating Problems

The objective of this section is to show how the sets of admissible vectors and optimizers of different linear optimization problems are related. The following notation is useful, to extract some coordinates of vectors.

Notation 1.12
Given a set $Q \subseteq \mathbb{R}^k$ and $1 \leq i \leq j \leq k$, we denote by

$$Q|_j^i$$

the set

$$\{(x_i, \ldots, x_j) \in \mathbb{R}^{j-i} : (x_1, \ldots, x_i, \ldots, x_j, \ldots, x_k) \in Q\}.$$

We omit i when $i = 1$ and j when $j = k$.

We start by showing that there is a canonical counterpart of every linear optimization problem. Moreover, their sets of admissible vectors and optimizers are the same modulo minor differences.

Proposition 1.1
Let $LC : \mathcal{L} \to \mathcal{C}$ be the map

$$LC\left(\begin{cases} \min_x cx \\ A'x \leq b' \\ A''x \geq b'' \\ A'''x = b''' \end{cases}\right) = \begin{cases} \max_y \begin{bmatrix} -c & c \end{bmatrix} y \\ \begin{bmatrix} A' & -A' \\ -A'' & A'' \\ A''' & -A''' \\ -A''' & A''' \end{bmatrix} y \leq \begin{bmatrix} b' \\ -b'' \\ b''' \\ -b''' \end{bmatrix} \\ y \geq 0. \end{cases}$$

Then,

$$X_P = X_{LC(P)}|_n - X_{LC(P)}|^{n+1} \text{ and } S_P = S_{LC(P)}|_n - S_{LC(P)}|^{n+1},$$

for each n-dimensional $P \in \mathcal{L}$. *Mutatis mutandis* for linear maximization problems.

Proof:

Assume that A' is an $m' \times n$-matrix, A'' is an $m'' \times n$-matrix and A''' is an $m''' \times n$-matrix.

(1) $X_P = X_{LC(P)}|_n - X_{LC(P)}|^{n+1}$. Given $x \in X_P$, let $y \in \mathbb{R}^{2n}$ be such that

$$(y_j, y_{n+j}) = \begin{cases} (x_j, 0) & \text{if } x_j \geq 0 \\ (0, -x_j) & \text{otherwise} \end{cases}$$

for each $j = 1, \ldots, n$. Observe that

$$y_j - y_{n+j} = x_j.$$

Hence,

$$\begin{bmatrix} A' & -A' \end{bmatrix} y = A' \left(\begin{bmatrix} y_1 \\ \vdots \\ y_n \end{bmatrix} - \begin{bmatrix} y_{n+1} \\ \vdots \\ y_{2n} \end{bmatrix} \right) = A'x \leq b'.$$

The other restrictions for $y \in X_{LC(P)}$ are proved in a similar way. Thus,

$$\begin{bmatrix} y_1 \\ \vdots \\ y_n \end{bmatrix} \in X_{LC(P)}|_n \quad \text{and} \quad \begin{bmatrix} y_{n+1} \\ \vdots \\ y_{2n} \end{bmatrix} \in X_{LC(P)}|^{n+1}.$$

So, $x \in X_{LC(P)}|_n - X_{LC(P)}|^{n+1}$, since

$$x = \begin{bmatrix} y_1 \\ \vdots \\ y_n \end{bmatrix} - \begin{bmatrix} y_{n+1} \\ \vdots \\ y_{2n} \end{bmatrix}.$$

For the other inclusion, let $y \in X_{LC(P)}$. Then, it is immediate that

$$\begin{bmatrix} y_1 \\ \vdots \\ y_n \end{bmatrix} - \begin{bmatrix} y_{n+1} \\ \vdots \\ y_{2n} \end{bmatrix}$$

is in X_P.

(2) $S_P = S_{LC(P)}|_n - S_{LC(P)}|^{n+1}$. Given $s \in S_P$ let $r \in \mathbb{R}^{2n}$ be such that

$$(r_j, r_{n+j}) = \begin{cases} (s_j, 0) & \text{if } s_j \geq 0 \\ (0, -s_j) & \text{otherwise} \end{cases}$$

for each $j = 1, \ldots, n$. Observe that

$$r_j - r_{n+j} = s_j.$$

Let $y \in X_{LC(P)}$. Then,

$$\begin{bmatrix} -c & c \end{bmatrix} r = -c \begin{bmatrix} r_1 \\ \vdots \\ r_n \end{bmatrix} + c \begin{bmatrix} r_{n+1} \\ \vdots \\ r_{2n} \end{bmatrix} = -cs$$

$$\geq$$

$$-c \left(\begin{bmatrix} y_1 \\ \vdots \\ y_n \end{bmatrix} - \begin{bmatrix} y_{n+1} \\ \vdots \\ y_{2n} \end{bmatrix} \right) = \begin{bmatrix} -c & c \end{bmatrix} y.$$

Thus, $r \in S_{LC(P)}$ and so $s \in S_{LC(P)}|_n - S_{LC(P)}|^{n+1}$. The proof of the other inclusion follows in a similar way. QED

Example 1.15
For instance,

$$LC \left(\begin{cases} \min_x \begin{bmatrix} -3 & -5 \end{bmatrix} x \\ \begin{bmatrix} -5 & 2 \\ 2 & -1 \end{bmatrix} x \leq \begin{bmatrix} 5 \\ 5 \end{bmatrix} \\ \begin{bmatrix} 6 & 1 \end{bmatrix} x \geq \begin{bmatrix} -7 \end{bmatrix} \end{cases} \right) =$$

$$= \begin{cases} \max_y \begin{bmatrix} 3 & 5 & -3 & -5 \end{bmatrix} y \\ \begin{bmatrix} -5 & 2 & 5 & -2 \\ 2 & -1 & -2 & 1 \\ -6 & -1 & 6 & 1 \end{bmatrix} y \leq \begin{bmatrix} 5 \\ 5 \\ 7 \end{bmatrix} \\ y \geq 0. \end{cases}$$

Next, our aim now is to relate a standard optimization problem with a canonical optimization problem in such a way that their sets of admissible vectors and optimizers are related, in a natural manner.

Proposition 1.2

Let $SC : \mathcal{S} \to \mathcal{C}$ be the map

$$SC \left(\left\{ \begin{array}{l} \min_{x} cx \\ Ax = b \\ x \geq 0 \end{array} \right. \right) = \left\{ \begin{array}{l} \max_{x} -cx \\ \left[\begin{array}{c} A \\ -A \end{array} \right] x \leq \left[\begin{array}{c} b \\ -b \end{array} \right] \\ x \geq 0. \end{array} \right.$$

Then, $X_P = X_{SC(P)}$ and $S_P = S_{SC(P)}$, for each $P \in \mathcal{S}$.

Proof:

Assume that A is an $m \times n$-matrix.

(1) $X_P = X_{SC(P)}$. It is enough to observe that, for every $i = 1, \ldots, m$,

$$\sum_{j=1}^{n} a_{ij}x_j = b_i \qquad \text{if and only if} \qquad \left\{ \begin{array}{l} \sum_{j=1}^{n} a_{ij}x_j \leq b_i \\ -\sum_{j=1}^{n} a_{ij}x_j \leq -b_i. \end{array} \right.$$

(2) $S_P = S_{SC(P)}$. Indeed,

$$cs \leq cx, \text{ for every } x \in X_P \quad \text{iff} \quad -cx \leq -cs, \text{ for every } x \in X_P$$

$$\text{iff} \quad -cx \leq -cs, \text{ for every } x \in X_{SC(P)}.$$

Hence, $s \in S_P$ if and only if $s \in S_{SC(P)}$.   QED

Example 1.16

For instance,

$$SC \left(\left\{ \begin{array}{l} \min_{x} \left[\begin{array}{cccc} -2 & -1 & 0 & 0 \end{array} \right] x \\ \left[\begin{array}{cccc} 3 & -1 & 0 & 1 \\ -1 & 3 & 1 & 0 \end{array} \right] x = \left[\begin{array}{c} 6 \\ 6 \end{array} \right] \\ x \geq 0 \end{array} \right. \right) =$$

$$
= \begin{cases} \max_{x} \begin{bmatrix} 2 & 1 & 0 & 0 \end{bmatrix} x \\[4pt] \begin{bmatrix} 3 & -1 & 0 & 1 \\ -1 & 3 & 1 & 0 \\ -3 & 1 & 0 & -1 \\ 1 & -3 & -1 & 0 \end{bmatrix} x \leq \begin{bmatrix} 6 \\ 6 \\ -6 \\ -6 \end{bmatrix} \\[4pt] x \geq 0. \end{cases}
$$

We now concentrate on obtaining a standard optimization problem from a canonical optimization problem.

Proposition 1.3
Let $CS : \mathcal{C} \to \mathcal{S}$ be the map

$$
CS \left(\begin{cases} \max_{x} cx \\ Ax \leq b \\ x \geq 0 \end{cases} \right) = \begin{cases} \min_{\bar{x}} \begin{bmatrix} -c & 0 \end{bmatrix} \bar{x} \\[4pt] \begin{bmatrix} A & I \end{bmatrix} \bar{x} = b \\[4pt] \bar{x} \geq 0. \end{cases}
$$

Then, $f : \mathbb{R}^n \to \mathbb{R}^{n+m}$ defined by

$$
f(x) = \begin{bmatrix} x \\ b - Ax \end{bmatrix}
$$

when restricted to X_P is a bijection between X_P and $X_{CS(P)}$ and when restricted to S_P is a bijection between S_P and $S_{CS(P)}$. Hence,

$$
X_P = X_{CS(P)}|_n \quad \text{and} \quad S_P = S_{CS(P)}|_n,
$$

for each n-dimensional $P \in \mathcal{C}$.

Proof:
Assume that A is an $m \times n$-matrix. It is immediate that f is injective. We now show that $f(X_P) = X_{CS(P)}$:
($\subseteq$) Given $x \in X_P$, it is easy to check that $f(x) \in X_{CS(P)}$. Indeed,

$$
\begin{bmatrix} A & I \end{bmatrix} f(x) = Ax + I(b - Ax) = b.
$$

Furthermore, $f(x) \geq 0$ since $x \geq 0$ and $Ax \leq b$, because $x \in X_P$.

($\supseteq$) Given $\overline{x} \in X_{CS(P)}$, we now show that there exists $x \in X_P$ such that $f(x) = \overline{x}$. Take x as the vector with the first n components of $\overline{x}$ and y the vector with the remaining components. Thus,

$$\overline{x} = \left[\begin{array}{c} x \\ y \end{array}\right].$$

Then, $x \geq 0$ because $\overline{x} \geq 0$. Furthermore, from

$$\left[\begin{array}{cc} A & I \end{array}\right] \overline{x} = b$$

it follows that $Ax + Iy = b$. Since $y \geq 0$, it follows that $Ax \leq b$. Hence, $x \in X_P$. Moreover, $y = b - Ax$. So,

$$\overline{x} = \left[\begin{array}{c} x \\ b - Ax \end{array}\right] = f(x).$$

Thus, f when restricted to X_P is a bijection between X_P and $X_{CS(P)}$.

(1) $X_P = X_{CS(P)}|_n$. Let $x \in X_P$. Then, $f(x) \in X_{CS(P)}$. Hence, the vector with the first n components of $f(x)$ (that is, x) is in $X_{CS(P)}|_n$. Let $x' \in X_{CS(P)}$. Then, there is $x \in X_P$ such that $f(x) = x'$. Then, the first n components of x' is in X_P since it is x.

(2) $S_P = S_{CS(P)}|_n$. Let $s \in S_P$. Let $x' \in X_{CS(P)}$. Then,

$$\left[\begin{array}{cc} -c & 0 \end{array}\right] f(s) = -cs \leq -cf^{-1}(x') = \left[\begin{array}{cc} -c & 0 \end{array}\right] x'.$$

Therefore, $f(s) \in S_{CS(P)}$. Hence, $s \in S_{CS(P)}|_n$. Let $s' \in S_{CS(P)}$ and $x \in X_P$. Then,

$$cf^{-1}(s') = -\left[\begin{array}{cc} -c & 0 \end{array}\right] s' \geq -\left[\begin{array}{cc} -c & 0 \end{array}\right] f(x) = cx.$$

So, $f^{-1}(s')$, that is, the vector with the first n components of s', is in S_P. QED

Example 1.17
For instance,

$$CS\left(\left\{\begin{array}{l} \max_{x} \left[\begin{array}{cc} 2 & 1 \end{array}\right] x \\[2mm] \left[\begin{array}{cc} 3 & -1 \\ -1 & 3 \end{array}\right] x \leq \left[\begin{array}{c} 6 \\ 6 \end{array}\right] \\[2mm] x \geq 0 \end{array}\right.\right) =$$

$$
\begin{cases}
\min_{x} \begin{bmatrix} -2 & -1 & 0 & 0 \end{bmatrix} x \\
= \begin{bmatrix} 3 & -1 & 1 & 0 \\ -1 & 3 & 0 & 1 \end{bmatrix} x = \begin{bmatrix} 6 \\ 6 \end{bmatrix} \\
x \geq 0.
\end{cases}
$$

The following result shows that deciding whether or not the set of admissible vectors of a canonical optimization problem is non-empty, is equivalent to deciding whether or not the set of optimizers of another canonical optimization problem is non-empty.

Proposition 1.4
Let P be the n-dimensional canonical optimization problem

$$
\begin{cases}
\max_{x} cx \\
Ax \leq b \\
x \geq 0
\end{cases}
$$

and P' the $n + m$-dimensional linear optimization problem

$$
\begin{cases}
\max_{x'} -ux' \\
\begin{bmatrix} A & -I \end{bmatrix} x' \leq b \\
x' \geq 0,
\end{cases}
$$

where $u \in \mathbb{R}^{n+m}$ is composed by n zeros followed by m ones. Then, $X_P \neq \emptyset$ if and only if there exists $s' \in S_{P'}$ such that $s'_j = 0$ for $j = n + 1, \ldots, n + m$.

Proof:
Observe that, for every $x' \in X_{P'}$,

$$
-ux' = -x'_{m+1} - \cdots - x'_{m+n} \leq 0.
$$

($\rightarrow$) Assume that $X_P \neq \emptyset$. Let $x \in X_P$. Observe that

$$
(x_1, \ldots, x_n, 0, \ldots, 0) \in X_{P'}.
$$

Therefore,
$$(x_1, \ldots, x_n, 0, \ldots, 0) \in S_{P'}$$
since $-u(x_1, \ldots, x_n, 0, \ldots, 0) = 0$.

($\leftarrow$) Suppose that $(s'_1, \ldots, s'_n, 0, \ldots, 0) \in S_{P'}$. Then, $A(s'_1, \ldots, s'_n) \leq b$ and $(s'_1, \ldots, s'_n) \geq 0$. Therefore, $(s'_1, \ldots, s'_n) \in X_P$.   QED

Finally, we present a map for getting a standard optimization problem with positive restriction vector from a given standard problem.

Exercise 1.4

Let $SP : \mathcal{S} \to \mathcal{S}$ be the map

$$SP\left(\begin{cases} \min_x cx \\ Ax = b \\ x \geq 0 \end{cases}\right) = \begin{cases} \min_y \begin{bmatrix} c & 0 \end{bmatrix} y \\ \begin{bmatrix} A^+ & 0 \\ -A^- & 0 \\ A^0 & 1 \\ 0 & 1 \end{bmatrix} y = \begin{bmatrix} b^+ \\ -b^- \\ 1 \\ 1 \end{bmatrix} \\ y \geq 0, \end{cases}$$

where

- A^+ and b^+ are obtained from A and b by removing the lines i such that $b_i \leq 0$, respectively;

- A^- and b^- are obtained from A and b by removing the lines i such that $b_i \geq 0$, respectively;

- A^0 is the matrix containing the lines of A that are neither in A^+ nor in A^-.

Show that, for each $P \in \mathcal{S}$,

$$X_{SP(P)} = \left\{ \begin{bmatrix} x \\ 1 \end{bmatrix} : x \in X_P \right\} \quad \text{and} \quad S_{SP(P)} = \left\{ \begin{bmatrix} s \\ 1 \end{bmatrix} : s \in S_P \right\}.$$

Exercise 1.5

Define a map for getting a canonical optimization problem with positive restriction vector from a given canonical problem in such a way that the sets of admissible vectors and optimizers are equivalent.

1.5   Solved Problems and Exercises

Problem 1.1 (Transportation Problem)
A company wants to minimize the weekly CO2 emission caused by the transportation of a given good between its factories and its distribution centers. Each distribution center needs at least a certain weekly amount of units of the good, that may vary from one center to another. Weekly, each factory produces several units of the product that may vary from one factory to another. Assume that the weekly CO2 emission for the transportation of a unit of good between a factory and a distribution center is known. Furthermore, the total weekly CO2 emission is proportional to the number of units transported. Model this problem as a linear optimization problem.

Solution:
Assume that

- the factories range from factory 1 to factory m;

- the distribution centers range from center 1 to center n;

- the weekly amount of units of the good required by distribution center j is at least v_j;

- the weekly amount of units of the good produced by factory i is u_i;

- the weekly CO2 emission for the transportation of a unit of the good between factory i and distribution center j is e_{ij}.

The goal is to minimize the total weekly CO2 emission produced by the transportation of the good between the factories and the distribution centers. The objective is attained by picking up the adequate number of units that should be transported from factory i to distribution center j that we represent by

$$y_{ij}.$$

Therefore, the total weekly CO2 emission is

$$\sum_{i=1}^{m}\sum_{j=1}^{n} e_{ij}y_{ij}.$$

We want to minimize this quantity taking into account the following conditions:

- $\sum_{i=1}^{m} y_{ij} \geq v_j$ for $j = 1, \ldots, n$; that is, the total weekly amount of units of the good transported to the distribution center j should be at least v_j;

- $\sum_{j=1}^{n} y_{ij} \leq u_i$ for $i = 1, \ldots, m$; that is, the total weekly amount of units of the good transported from factory i is at most u_i.

Thus, the problem is represented as follows:

$$\begin{cases} \displaystyle\min_{(y_{11},\ldots,y_{mn})} \sum_{i=1}^{m} \sum_{j=1}^{n} e_{ij} y_{ij} & \\[2ex] \displaystyle\sum_{j=1}^{n} y_{ij} \leq u_i & \text{for } i = 1, \ldots, m \\[2ex] \displaystyle\sum_{i=1}^{m} y_{ij} \geq v_j & \text{for } j = 1, \ldots, n \\[2ex] (y_{11}, \ldots, y_{mn}) \geq 0. \end{cases}$$

To present the problem in matricial form consider a bijection:

$$\beta : \{1, \ldots, mn\} \to \{1, \ldots, m\} \times \{1, \ldots, n\}.$$

Hence, the problem is:

$$\begin{cases} \displaystyle\min_{x} \begin{bmatrix} e_{\beta(1)} & \cdots & e_{\beta(mn)} \end{bmatrix} x & \\[3ex] \begin{bmatrix} a_{11} & \cdots & a_{1\,mn} \\ \vdots & \ddots & \vdots \\ a_{m1} & \cdots & a_{m\,mn} \end{bmatrix} x \leq \begin{bmatrix} u_1 \\ \vdots \\ u_m \end{bmatrix} & \\[4ex] \begin{bmatrix} a'_{11} & \cdots & a'_{1\,mn} \\ \vdots & \ddots & \vdots \\ a'_{n1} & \cdots & a'_{n\,mn} \end{bmatrix} x \geq \begin{bmatrix} v_1 \\ \vdots \\ v_n \end{bmatrix}, \end{cases}$$

where

$$a_{ij} = \begin{cases} 1 & \text{if } \beta(j)_1 = i \\ 0 & \text{otherwise} \end{cases}$$

and

$$a'_{ij} = \begin{cases} 1 & \text{if } \beta(j)_2 = i \\ 0 & \text{otherwise} \end{cases}$$

and x is $(y_{\beta(1)}, \ldots, y_{\beta(mn)})$.   ◁

Problem 1.2
Consider the following linear optimization problem P:

$$\begin{cases} \max_{x} \; 2x_1 + x_2 \\[4pt] -x_1 - x_2 \leq -1 \\[4pt] x_1 + x_2 \leq 3 \\[4pt] x_1 - x_2 \leq 1 \\[4pt] -x_1 + x_2 \leq 1 \\[4pt] x \geq 0. \end{cases}$$

(1) Present the problem in matricial form.

(2) Define and provide a graphical presentation of X_P. Verify whether or not $(2,1)$, $(0,2)$ and $(1,1)$ are in X_P.

(3) Verify whether or not $(0,2)$ and $(1,1)$ are in S_P.

(4) Find the standard optimization problem corresponding to P according to the reduction $CS : \mathcal{C} \to \mathcal{S}$.

(5) Find the canonical optimization problem corresponding to $CS(P)$ according to the reduction $SC : \mathcal{S} \to \mathcal{C}$.

(6) Is $CS : \mathcal{C} \to \mathcal{S}$ the inverse of $SC : \mathcal{S} \to \mathcal{C}$?

Solution:
(1) The matricial form is as follows:

$$\begin{cases} \max_{x} \begin{bmatrix} 2 & 1 \end{bmatrix} x \\[6pt] \begin{bmatrix} -1 & -1 \\ 1 & 1 \\ 1 & -1 \\ -1 & 1 \end{bmatrix} x \leq \begin{bmatrix} -1 \\ 3 \\ 1 \\ 1 \end{bmatrix} \\[6pt] x \geq 0. \end{cases}$$

(2) The set X_P of admissible vectors is:

$$\{(x_1, x_2) \in \mathbb{R}^2 : -x_1 - x_2 \leq -1, x_1 + x_2 \leq 3, x_1 - x_2 \leq 1, -x_1 + x_2 \leq 1, x_1, x_2 \geq 0\}.$$

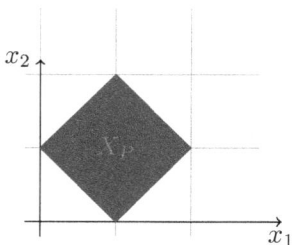

Figure 1.2: Set of admissible vectors.

Vector $(2,1)$ is in X_P since

$$\begin{bmatrix} -1 & -1 \\ 1 & 1 \\ 1 & -1 \\ -1 & 1 \end{bmatrix} \begin{bmatrix} 2 \\ 1 \end{bmatrix} = \begin{bmatrix} -3 \\ 3 \\ 1 \\ -1 \end{bmatrix} \leq \begin{bmatrix} -1 \\ 3 \\ 1 \\ 1 \end{bmatrix}$$

and

$$\begin{bmatrix} 2 \\ 1 \end{bmatrix} \geq \begin{bmatrix} 0 \\ 0 \end{bmatrix}.$$

Vector $(0,2)$ is not in X_P since

$$\begin{bmatrix} -1 & -1 \\ 1 & 1 \\ 1 & -1 \\ -1 & 1 \end{bmatrix} \begin{bmatrix} 0 \\ 2 \end{bmatrix} = \begin{bmatrix} -2 \\ 2 \\ -2 \\ 2 \end{bmatrix} \nleq \begin{bmatrix} -1 \\ 3 \\ 1 \\ 1 \end{bmatrix}.$$

Vector $(1,1)$ is in X_P since

$$\begin{bmatrix} -1 & -1 \\ 1 & 1 \\ 1 & -1 \\ -1 & 1 \end{bmatrix} \begin{bmatrix} 1 \\ 1 \end{bmatrix} = \begin{bmatrix} -2 \\ 2 \\ 0 \\ 0 \end{bmatrix} \leq \begin{bmatrix} -1 \\ 3 \\ 1 \\ 1 \end{bmatrix}$$

and

$$\begin{bmatrix} 1 \\ 1 \end{bmatrix} \geq \begin{bmatrix} 0 \\ 0 \end{bmatrix}.$$

(3) Vector $(0,2)$ is not in S_P since it is not in X_P.

Vector $(1,1)$ is not in S_P since $(2,1)$ is in X_P and

$$\begin{bmatrix} 2 & 1 \end{bmatrix} \begin{bmatrix} 2 \\ 1 \end{bmatrix} = 5$$

and

$$[\ 2 \ \ 1\ ] \begin{bmatrix} 1 \\ 1 \end{bmatrix} = 3.$$

So, $(1,1)$ is not a maximizer of P.

(4) The standard optimization problem $CS(P)$ corresponding to P according to the reduction $CS : \mathcal{C} \to \mathcal{S}$ is as follows:

$$\begin{cases} \min\limits_{x} \ -2x_1 - x_2 \\[4pt] -x_1 - x_2 + x_3 = -1 \\[4pt] x_1 + x_2 + x_4 = 3 \\[4pt] x_1 - x_2 + x_5 = 1 \\[4pt] -x_1 + x_2 + x_6 = 1 \\[4pt] x \geq 0 \end{cases}$$

or in matricial form

$$\begin{cases} \min\limits_{x} \begin{bmatrix} -2 & -1 & 0 & 0 & 0 & 0 \end{bmatrix} x \\[4pt] \begin{bmatrix} -1 & -1 & 1 & 0 & 0 & 0 \\ 1 & 1 & 0 & 1 & 0 & 0 \\ 1 & -1 & 0 & 0 & 1 & 0 \\ -1 & 1 & 0 & 0 & 0 & 1 \end{bmatrix} x = \begin{bmatrix} -1 \\ 3 \\ 1 \\ 1 \end{bmatrix} \\[4pt] x \geq 0. \end{cases}$$

(5) The canonical optimization problem $SC(CS(P))$ corresponding to $CS(P)$

according to the reduction $SC : \mathcal{S} \to \mathcal{C}$ is as follows:

$$
\begin{cases}
\max_{x} \ 2x_1 + x_2 \\
-x_1 - x_2 + x_3 \le -1 \\
x_1 + x_2 + x_4 \le 3 \\
x_1 - x_2 + x_5 \le 1 \\
-x_1 + x_2 + x_6 \le 1 \\
+x_1 + x_2 - x_3 \le 1 \\
-x_1 - x_2 - x_4 \le -3 \\
-x_1 + x_2 - x_5 \le -1 \\
+x_1 - x_2 - x_6 \le -1 \\
x_1, \ldots, x_6 \ge 0
\end{cases}
$$

or in matricial form:

$$
\begin{cases}
\max_{x} \begin{bmatrix} 2 & 1 & 0 & 0 & 0 & 0 \end{bmatrix} x \\[2mm]
\begin{bmatrix}
-1 & -1 & 1 & 0 & 0 & 0 \\
1 & 1 & 0 & 1 & 0 & 0 \\
1 & -1 & 0 & 0 & 1 & 0 \\
-1 & 1 & 0 & 0 & 0 & 1 \\
1 & 1 & -1 & 0 & 0 & 0 \\
-1 & -1 & 0 & -1 & 0 & 0 \\
-1 & 1 & 0 & 0 & -1 & 0 \\
1 & -1 & 0 & 0 & 0 & -1
\end{bmatrix} x \le
\begin{bmatrix}
-1 \\ 3 \\ 1 \\ 1 \\ 1 \\ -3 \\ -1 \\ -1
\end{bmatrix} \\[2mm]
x \ge 0.
\end{cases}
$$

(6) The map $CS : \mathcal{C} \to \mathcal{S}$ is inverse of the map $SC : \mathcal{S} \to \mathcal{C}$ if and only if

- $CS \circ SC = \mathrm{id}_{\mathcal{S}}$; that is, $\mathcal{S} \xrightarrow{SC} \mathcal{C} \xrightarrow{CS} \mathcal{S}$ is equal to $\mathcal{S} \xrightarrow{\mathrm{id}_{\mathcal{S}}} \mathcal{S}$;

- $SC \circ CS = \mathrm{id}_{\mathcal{C}}$; that is, $\mathcal{C} \xrightarrow{CS} \mathcal{S} \xrightarrow{SC} \mathcal{C}$ is equal to $\mathcal{C} \xrightarrow{\mathrm{id}_{\mathcal{C}}} \mathcal{C}$.

Hence, the map CS is not the inverse of the map SC since

$$SC(CS(P)) \neq P,$$

as we have seen before.   ◁

Exercise 1.6

Consider the following linear optimization problem P:

$$\begin{cases} \min_{x} \ -3x_1 - 4x_2 \\ -8x_1 - 3x_2 \geq -5 \\ -6x_1 + 4x_2 \geq 5 \\ 2x_1 - x_2 \leq 2. \end{cases}$$

Present the problem in matricial form. Obtain the corresponding canonical optimization problem given by the map $LC : \mathcal{L} \to \mathcal{C}$. Relate their sets of admissible vectors and optimizers.

Exercise 1.7

Consider the following canonical optimization problem P:

$$\begin{cases} \max_{x} \ 2x_1 + x_2 \\ x_1 \leq 5 \\ 4x_1 + x_2 \leq 25 \\ x \geq 0. \end{cases}$$

(1) Present the problem in matricial form.

(2) Define and provide a graphical presentation of X_P. Verify whether or not $(5, 5)$ and $(7, 2)$ are in X_P.

(3) Verify whether or not $(7, 2)$ is in S_P.

(4) Find the standard optimization problem corresponding to P according to the reduction $CS : \mathcal{C} \to \mathcal{S}$. Present $X_{CS(P)}$ and $S_{CS(P)}$.

(5) Find the canonical optimization problem corresponding to $CS(P)$ according to the reduction $SC : \mathcal{S} \to \mathcal{C}$. Present $X_{SC(CS(P))}$ and $S_{SC(CS(P))}$.

1.6   Relevant Background

We start by introducing some basic algebraic concepts and notations, namely about groups, fields and vector spaces. The interested reader can consult [38, 15, 7, 47, 59].

Definition 1.10
A *group* is a pair
$$(G, +),$$
where

- G is a non-empty set;

- $+ : G^2 \to G$ is a map;

such that

- $x + (y + z) = (x + y) + z$;

- there exists $0 \in G$ such that $0 + x = x + 0 = x$ (*identity*);

- for every $x \in G$ there is $y \in G$ with $x + y = 0$.

The group is *Abelian* whenever $x + y = y + x$ for every $x, y \in G$.

Example 1.18
The pair
$$(\mathbb{R}, +),$$
where $+$ is the sum of real numbers, is an Abelian group. On the other hand, $(\mathbb{N}, +)$, where $+$ is the sum of natural numbers, is not a group.

Example 1.19
Let $(G, +)$ be a group. Then, the pair
$$(G^n, +),$$
where

- $G^n = \{(x_1, \ldots, x_n) : x_j \in G, j = 1, \ldots, n\}$;

- $(x_1, \ldots, x_n) + (y_1, \ldots, y_n) = (x_1 + y_1, \ldots, x_n + y_n)$;

is a group. Observe that the identity is the tuple $(0, \ldots, 0)$.

Remark 1.3

Given a group $(G, +)$ and $x \in G$, we denote by

$$-x$$

the unique element of G such that

$$x + (-x) = 0.$$

Moreover, we may write

$$x - y$$

instead of

$$x + (-y)$$

for every $x, y \in G$.

Definition 1.11

A *field* is a tuple

$$(K, +, \times),$$

where

- K is a non-empty set;

- $+, \times : K^2 \to K$;

such that

- $(K, +)$ is an Abelian group with identity 0;

- $x \times (y \times z) = (x \times y) \times z$;

- $x \times y = y \times x$;

- $x \times (y + z) = (x \times y) + (x \times z)$;

- there is $1 \in K$ such that $1 \times x = x$ (*multiplicative identity*);

- $0 \neq 1$;

- if $x \neq 0$ then there is $w \in K$ such that $x \times w = 1$;

for every $x, y, z \in K$.

Example 1.20
The tuple
$$(\mathbb{Q}, +, \times),$$
where $+$ and $\times$ are the usual operations over rational numbers, is a field. Furthermore, the tuple
$$(\mathbb{R}, +, \times),$$
where $+$ and $\times$ are the usual operations over real numbers, is a field.

Notation 1.13
In the sequel, we may denote a field $(K, +, \times)$ by K.

We are ready to present the essential concept of vector space.

Definition 1.12
A *vector space* over a field K is an Abelian group $(V, +)$ with a map
$$(\alpha, x) \mapsto \alpha x : K \times V \to V$$
satisfying the following properties:

- $(\alpha \mu) x = \alpha(\mu x)$;

- $(\alpha + \mu) x = \alpha x + \mu x$;

- $\alpha(x + y) = \alpha x + \alpha y$;

- $1x = x$, where 1 is the multiplicative identity of K.

The elements of V and the elements of K are called *vectors* and *scalars*, respectively.

Example 1.21
Let K be a field, $\{v\}$ a singleton set and $+ : \{v\}^2 \to \{v\}$ such that $v + v = v$. Then, $(\{v\}, +)$ and the map $(\alpha, v) \mapsto v : K \times \{v\} \to \{v\}$ is a vector space over K.

Notation 1.14
Consider the vector space introduced in Example 1.21. We start by observing that v is the identity of the group and so is usually denoted by 0. Furthermore, we denote such a vector space by
$$\{0\}_K$$

and refer to it as the *zero vector space* over K.

Definition 1.13
The *vector space induced by a field K and $n \in \mathbb{N}^+$*, denoted by

$$K^n,$$

is the vector space $(K^n, +)$ over K with the map

$$(\alpha, (\beta_1, \ldots, \beta_n)) \mapsto (\alpha \times \beta_1, \ldots, \alpha \times \beta_n),$$

where $(\alpha_1, \ldots, \alpha_n) + (\beta_1, \ldots, \beta_n) = (\alpha_1 + \beta_1, \ldots, \alpha_n + \beta_n)$. The

$$(0, \ldots, 0)$$

element of K^n, denoted by 0, is called the *null vector*.

Example 1.22
We denote by

$$\mathbb{R}^n,$$

where $n \in \mathbb{N}^+$, the vector space induced by $\mathbb{R}$ and n.

We now introduce some notions and results related to subspaces.

Definition 1.14
A *subspace S* of a vector space V over a field K is a subset of V containing 0, closed under vector addition and multiplication by scalar.

Proposition 1.5
A subspace of a vector space V over a field K with the multiplication by scalar and addition induced by V in S is a vector space over K.

Example 1.23
The smallest subspace of a vector space V over a field K is $\{0\}_K$. That is,

$$\{0\}_K \subseteq V_1$$

for every subspace V_1 of V.

Definition 1.15
Let V be a vector space over a field K and $V_1, V_2 \subseteq V$. Then,

$$V_1 + V_2,$$

called the *sum* of V_1 and V_2, is the set

$$\{v^1 + v^2 : v^1 \in V_1 \text{ and } v^2 \in V_2\}.$$

Proposition 1.6
The class of all subspaces of a vector space over a field K is closed under intersections and finite sums. On the other hand, it is not closed under unions except in trivial case.

We now introduce the important concept of span of a set of vectors.

Definition 1.16
Let V be a vector space over a field K and $U \subseteq V$. Then, the set

$$\mathrm{span}_V(U),$$

the *span* of U in V, is

$$\bigcap_{\{V_1 : V_1 \text{ is a subspace of } V \text{ and } U \subseteq V_1\}} V_1.$$

The following result is a direct consequence of Proposition 1.6.

Proposition 1.7
Let V be a vector space over a field K and $U \subseteq V$. Then, $\mathrm{span}_V(U)$ is a subspace of V.

Example 1.24
Let V be a vector space over a field K. Note that

$$\mathrm{span}_V(\emptyset) = \{0\}_K,$$

since $\{0\}_K$ is a subspace of V containing $\emptyset$ and

$$\{0\}_K$$

is the smallest subspace of V, see Example 1.23.

Proposition 1.8
Let V be a vector space over a field K and $V_1, V_2 \subseteq V$. Then,

$$\mathrm{span}_V(V_1 \cup V_2) = V_1 + V_2.$$

The following result provides another characterization of span of a subset in V assuming the convention that

$$\sum_{j=1}^{0} \alpha_j u^j$$

is the 0 vector in V.

Proposition 1.9
Let V be a vector space over a field K and $U \subseteq V$. Then,

$$\mathrm{span}_V(U) = \left\{ \sum_{j=1}^{k} \alpha_j u^j : k \in \mathbb{N}, \alpha_j \in K, u^j \in U \right\}.$$

Hence, the span of a set U is seen as the set of all *linear combination of vectors* in U.

We now discuss finite-dimensional vector spaces, and begin by stating their definition.

Definition 1.17
A vector space V over a field K is *finite-dimensional* whenever there is a finite set $U \subseteq V$ such that $\mathrm{span}_V(U) = V$.

Remark 1.4
From now on, when referring to vector spaces we mean finite-dimensional vector spaces.

Definition 1.18
A finite set U is *linearly independent* in a vector space V over a field K whenever $U \subseteq V$ and for any $u^1, \ldots, u^n \in U$ and $\alpha_1, \ldots, \alpha_n \in K$, if

$$\alpha_1 u^1 + \cdots + \alpha_n u^n = 0$$

then $\alpha_1, \ldots, \alpha_n = 0$. Otherwise, the set is *linearly dependent* in V.

Example 1.25

The sets

$$\{(1,0),(0,1)\}   \text{and}   \{(2,2),(-3,2)\}$$

are linearly independent in $\mathbb{R}^2$. On the other hand, the set

$$\{(1,1),(2,2)\}$$

is not linearly independent in $\mathbb{R}^2$. Indeed, taking $\alpha_1 = 2$ and $\alpha_2 = -1$ we get
$2(1,1) - (2,2) = 0$.

Example 1.26

The set

$$\emptyset$$

is linearly independent in every vector space.

Proposition 1.10

Let V be a vector space over a field K and $U_1 \subseteq U_2 \subseteq V$. Then, U_1 is linearly
independent in V whenever U_2 is linearly independent in V.

Definition 1.19

Let V be a vector space over a field K. A finite set B is a *basis* of V whenever

$$V = \text{span}_V(B)$$

and B is a linearly independent set in V.

Example 1.27

The set $\{(1,0),(0,1)\}$ is a basis of $\mathbb{R}^2$.

Definition 1.20

The *standard or canonical basis* of a vector space K^n induced by a field K is
composed by the vectors

$$e^1 = (1,0,\ldots,0), e^2 = (0,1,0,\ldots,0),\ldots, e^n = (0,\ldots,0,1).$$

Example 1.28

Let K be a field. Observe that

$$\emptyset \text{ is a basis of } \{0\}_K$$

taking into account Example 1.24 and Example 1.26.

An important consequence of the following result is that every vector space has a basis.

Proposition 1.11

Let V be a vector space over a field K and $U \subseteq V$ a finite set with $\mathrm{span}_V(U) = V$. Then, there is a linearly independent set $B \subseteq U$ in V such that $\mathrm{span}_V(B) = V$.

Proposition 1.12

Every vector space over a field K has a basis. Moreover, all bases have the same number of elements.

We now define maps between vector spaces over the same field.

Definition 1.21

Let V and W be vector spaces over the same field K. We say that a map $h : V \to W$ is *linear* whenever the following properties hold:

- $h(x + y) = h(x) + h(y)$;

- $h(\alpha x) = \alpha h(x)$;

for $x, y \in V$ and $\alpha \in K$.

Exercise 1.8

Show that h is linear if and only if $h(\alpha x + \beta y) = \alpha h(x) + \beta h(y)$.

Example 1.29

Let W be a vector space over K. Then, $h : \{0\}_K \to W$ such that $h(0) = 0$ is a linear map.

Definition 1.22

Let V and W be vector spaces over the same field K with bases $v_1, \ldots, v_n$ and $w_1, \ldots, w_m$, respectively. The $m \times n$-*matrix*, or simply the *matrix*, induced by the linear map $h : V \to W$ with respect to the given bases, denoted by

$$M(h, (v_1, \ldots, v_n), (w_1, \ldots, w_m)),$$

is an m-by-n tuple

$$(a_{ij}) \in K^{m \times n}$$

such that
$$h(v_j) = a_{1j}w_1 + \cdots + a_{mj}w_m.$$

Notation 1.15
When either V or W is $\{0\}_K$, the matrix induced by the linear map $h : V \to W$ is called the *empty matrix*, denoted by

$$[\,].$$

Notation 1.16
When V and W are the vector spaces $\mathbb{R}^n$ and $\mathbb{R}^m$ over $\mathbb{R}$, respectively, and $v_1, \ldots, v_n$ and $w_1, \ldots, w_m$ are their standard or canonical bases, we may write

$$M(h)$$

for $M(h, (v_1, \ldots, v_n), (w_1, \ldots, w_m))$.

Example 1.30
Let $h : \mathbb{R}^2 \to \mathbb{R}^3$ be such that $h(x_1, x_2) = (x_1 - 3x_2, x_1, x_2)$. Then,

$$M(h)$$

is the matrix
$$\begin{bmatrix} 1 & -3 \\ 1 & 0 \\ 0 & 1 \end{bmatrix}.$$

Remark 1.5
In the sequel, we may refer to a vector $(x_1, \ldots, x_n) \in \mathbb{R}^n$ as

$$\begin{bmatrix} x_1 \\ \vdots \\ x_n \end{bmatrix}.$$

Remark 1.6
In the sequel, we may present a matrix without referring to the underlying linear map.

Definition 1.23
A *submatrix* of a matrix is obtained by deleting any collection of rows and/or columns of the matrix.

Chapter 2

Optimizers

This chapter investigates the existence of optimizers for linear optimization problems using topological techniques. In Section 2.1, we introduce the notion of interior and boundary of a canonical optimization problem and show that these concepts coincide with the topological ones with the same name. The main result of the section states that no maximizer is an interior point when $c \neq 0$. In Section 2.2, we prove sufficient conditions for the existence of maximizers after showing that the sets of admissible vectors and maximizers are closed. We also provide in Section 2.4, a *modicum* of relevant topological notions and results.

2.1 Boundary

The first insight provided by the topological analysis of the linear optimization problem is the characterization of the subset of the set of admissible vectors where we should look for the optimizers. We start by defining the interior and the boundary of the set of admissible vectors. It turns out that these notions agree with the corresponding topological notions.

Definition 2.1
The *interior* of the set of admissible vectors of a canonical n-dimensional optimization problem P

$$\begin{cases} \max_{x} cx \\ Ax \leq b \\ x \geq 0, \end{cases}$$

denoted by
$$X_P^\circ,$$
is the set $\{x \in \mathbb{R}^n : Ax < b \text{ and } x > 0\}$. The *boundary* of X_P, denoted by
$$\partial X_P,$$
is the set $X_P \setminus X_P^\circ$.

Example 2.1
Recall Example 1.12 and Example 1.14. Note that

- $X_P = \{(x_1, x_2) \in \mathbb{R}^2 : 3x_1 - x_2 \leq 6, -x_1 + 3x_2 \leq 6, x_1 \geq 0, x_2 \geq 0\}$;

- $X_P^\circ = \{(x_1, x_2) \in X_P : 3x_1 - x_2 < 6, -x_1 + 3x_2 < 6, x_1 > 0, x_2 > 0\}$;

- ∂X_P is the set

$$\{(x_1, x_2) \in X_P : 3x_1 - x_2 = 6\}$$
$$\cup \{(x_1, x_2) \in X_P : -x_1 + 3x_2 = 6\}$$
$$\cup \{(x_1, x_2) \in X_P : x_1 = 0\}$$
$$\cup \{(x_1, x_2) \in X_P : x_2 = 0\}.$$

Then, $(1, 1) \in X_P^\circ$ and $(3, 3) \in \partial X_P$. Moreover, $(3, 3) \notin X_P^\circ$ and $(1, 1) \notin \partial X_P$.

We are now ready to prove that the interior of the set of admissible vectors of a linear optimization problem coincides with the topological interior (see Section 2.4). Recall that each line of matrix A is non-null (see Remark 1.2). The first step is to characterize, from a topological point of view, the constraint map (as well as the objective map).

Proposition 2.1
Every linear map from $\mathbb{R}^n$ to $\mathbb{R}^m$ is continuous.

Proof:
Observe that $|x_j| \leq \|x\|$, for every $j = 1, \ldots, n$, by Proposition 2.12. Therefore,

$$\sum_{j=1}^{n} |x_j| \leq n\|x\| \qquad (\dagger).$$

Let $f : \mathbb{R}^n \to \mathbb{R}^m$ be a linear map and $\mu = \max\{\|f(e^j)\| : j = 1, \ldots, n\}$ (see Definition 1.20). Then,

$$
\begin{aligned}
\|f(x)\| &= \left\| f\left(\sum_{j=1}^{n} x_j e^j\right) \right\| \\
&= \left\| \sum_{j=1}^{n} x_j f(e^j) \right\| && \text{by linearity of } f \text{ (see Exercise 1.8)} \\
&\leq \sum_{j=1}^{n} \|x_j f(e^j)\| && \text{by triangular inequality} \\
&= \sum_{j=1}^{n} |x_j| \|f(e^j)\| && \text{by positive homogeneity} \\
&\leq \mu \sum_{j=1}^{n} |x_j| && \text{by definition of } \mu \\
&\leq \mu n \|x\| && \text{by (†);}
\end{aligned}
$$

that is,
$$
\|f(x)\| \leq \mu n \|x\|, \qquad (\ddagger).
$$

Finally, we show that f is continuous. Consider two cases:

(1) $\mu = 0$. Then, $f(x) = 0$ for every x (by positive definiteness). It is immediate that f is continuous.

(2) $\mu \neq 0$. Let $\{x^k\}_{k \in \mathbb{N}}$ be a sequence converging to x. We show that $\{f(x^k)\}_{k \in \mathbb{N}}$ is a sequence converging to $f(x)$. Take any $\delta > 0$. Note that

$$
\frac{\delta}{\mu n} \in \mathbb{R}^+.
$$

Since $\{x^k\}_{k \in \mathbb{N}}$ converges to x, there exists j such that

$$
\|x^m - x\| < \frac{\delta}{\mu n}
$$

for every $m \geq j$. Hence, for every $m \geq j$, it follows that

$$
\begin{aligned}
\|f(x^m) - f(x)\| &= \|f(x^m - x)\| && \text{by linearity of } f \\
&\leq \mu n \|x^m - x\| && \text{by (‡)} \\
&< \delta.
\end{aligned}
$$

Thus, $\{f(x^k)\}_{k \in \mathbb{N}}$ converges to $f(x)$. QED

Hence, we have the following corollary (recall Notation 1.8).

Proposition 2.2
The objective and the constraint maps of a (pure) canonical optimization problem are continuous.

Proof:
Immediate by Proposition 2.1 taking into account that, by definition, the objective and the constraint maps of a linear optimization problem are linear maps.   QED

We now show that the interior of a canonical optimization problem coincides with its topological counterpart (see Notation 2.1).

Proposition 2.3
Consider a canonical linear optimization problem P. Then,

$$X_P^\circ = \mathsf{int}(X_P).$$

Proof:
Assume that P is in pure form and that A is an $m \times n$ matrix.

($\subseteq$) Let $x \in X_P^\circ$. Then,

$$\underline{A}x < \underline{b}.$$

We must find $\varepsilon > 0$ such that

$$B_\varepsilon(x) \subset X_P.$$

Let

$$\delta = \min\{(\underline{b}_i - (\underline{A}x)_i) : 1 \leq i \leq m + n\}.$$

Note that $\delta > 0$. Since $x \mapsto \underline{A}x$ is continuous (by Proposition 2.2), then, by Proposition 2.23, there exists $\varepsilon > 0$ such that

$$\text{if } \|y - x\| < \varepsilon \text{ then } \|\underline{A}y - \underline{A}x\| < \delta.$$

In other words, let $\varepsilon > 0$ be such that

$$\text{if } y \in B_\varepsilon(x) \text{ then } \|\underline{A}y - \underline{A}x\| < \delta.$$

We now show that $B_\varepsilon(x) \subseteq X_P^\circ$. Let $y \in B_\varepsilon(x)$. Then, from $\|\underline{A}y - \underline{A}x\| < \delta$, using Proposition 2.12, it follows that

$$|(\underline{A}y)_i - (\underline{A}x)_i| < \delta$$

for each $i = 1, \ldots, m + n$. We prove that

$$(\underline{A}y)_i < \underline{b}_i$$

for every $i = 1, \ldots, m + n$. There are two cases to consider:

(1) $(\underline{A}y)_i \leq (\underline{A}x)_i$. Then, we have $(\underline{A}y)_i < \underline{b}_i$ since $(\underline{A}x)_i < b_i$.

(2) $(\underline{A}y)_i > (\underline{A}x)_i$. Hence,

$$(\underline{A}y)_i - (\underline{A}x)_i = |(\underline{A}y)_i - (\underline{A}x)_i| < \delta.$$

Therefore,

$$(\underline{A}y)_i < \delta + (\underline{A}x)_i \leq \underline{b}_i - (\underline{A}x)_i + (\underline{A}x)_i = \underline{b}_i.$$

Therefore, $y \in X_P^\circ$. Since $X_P^\circ \subset X_P$, then $B_\varepsilon(x) \subset X_P$. Hence, $x \in \mathsf{int}(X_P)$.

($\supseteq$) Assume that $x \in \mathsf{int}(X_P)$. Then, let $\varepsilon > 0$ be such that $B_\varepsilon(x) \subset X_P$. Take $v \in \mathbb{R}^n$ such that

- $v_j \neq 0$ for $j = 1, \ldots, n$;

- $\sum_{j=1}^n a_{ij} v_j \neq 0$ for $i = 1, \ldots, m$ (see Remark 1.2);

- $\|v\| < \varepsilon$.

Then, $x + v \in B_\varepsilon(x)$ and $x - v \in B_\varepsilon(x)$ by the last constraint on v. So, by hypothesis, $x + v \in X_P$ and $x - v \in X_P$. We now prove that

$$x \in X_P^\circ.$$

Indeed:

(1) $x_j > 0$ for $j = 1, \ldots, n$. Observe that

$$\begin{cases} (\dagger) & x_j + v_j \geq 0 \\ (\ddagger) & x_j - v_j \geq 0. \end{cases}$$

By the first constraint on v, $v_j \neq 0$. So, we have two cases:

(a) $v_j < 0$. Then, by $(\dagger)$, $x_j > 0$;

(b) $v_j > 0$. Then, by $(\ddagger)$, $x_j > 0$.

(2) $(Ax)_i < b_i$, for $i = 1, \ldots, m$ using the second constraint on v and reasoning as in (1). Observe that $(Ax)_i + (Av)_i \leq b_i$ and $(Ax)_i - (Av)_i \leq b_i$.   QED

Exercise 2.1

Let P be a canonical optimization problem. Show that $\partial X_P = \mathsf{bn}(X_P)$.

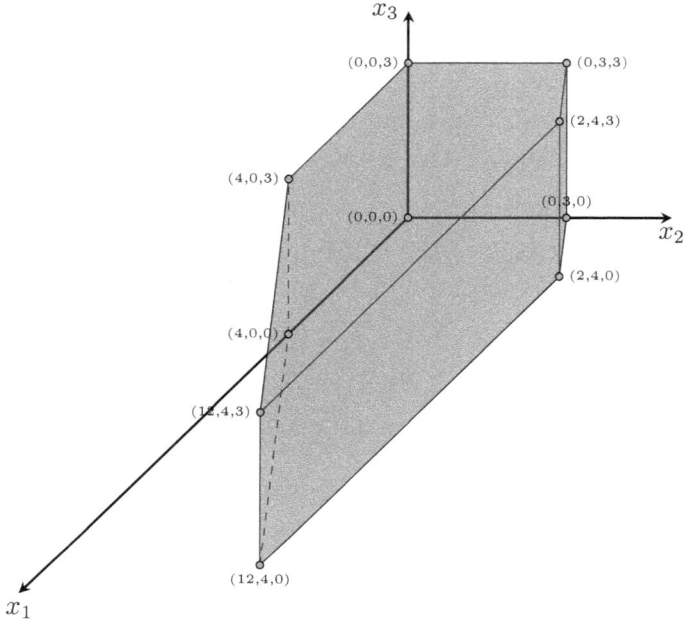

Figure 2.1: Set of admissible vectors of problem in Example 2.2

Example 2.2
Consider the following linear optimization problem P

$$\begin{cases} \max_{(x_1,x_2,x_3)} \quad 3x_1 + 3x_2 + 2x_3 \\[1mm] x_2 \leq 4 \\ x_3 \leq 3 \\ 2x_2 - x_1 \leq 6 \\ -2x_2 + x_1 \leq 4 \\ x_1, x_2, x_3 \geq 0 \end{cases}$$

with set of admissible vectors depicted in Figure 2.1. We now present examples of vectors in the interior and in the boundary of X_P. We start by showing that

$$(2, 2, 2) \in X_P^\circ.$$

Observe that X_P° is the set

$$\{(x_1, x_2, x_3) \in (\mathbb{R}^+)^3 : x_2 < 4, x_3 < 3, 2x_2 - x_1 < 6, -2x_2 + x_1 < 4\}.$$

Hence, $(2, 2, 2) \in X_P^\circ$.

Another way to show that $(2, 2, 2) \in X_P^\circ$ is to prove that

$$(2, 2, 2) \in \text{int}(X_P)$$

and then use Proposition 2.3. Hence, we must find a real number $\varepsilon > 0$ such that $B_\varepsilon(2, 2, 2) \subset X_P$. Take ε equal to $\frac{1}{2}$. It remains to show that $B_{\frac{1}{2}}(2, 2, 2) \subseteq X_P$. Let $y \in B_{\frac{1}{2}}(2, 2, 2)$. Then,

$$\|(2, 2, 2) - (y_1, y_2, y_3)\| < \frac{1}{2}.$$

That is,

$$(2 - y_1)^2 + (2 - y_2)^2 + (2 - y_3)^2 < \frac{1}{4}.$$

Thus, $(2 - y_1)^2 < \frac{1}{4}$, $(2 - y_2)^2 < \frac{1}{4}$ and $(2 - y_3)^2 < \frac{1}{4}$. Therefore,

$$|2 - y_1| < \frac{1}{2}, \quad |2 - y_2| < \frac{1}{2} \quad \text{and} \quad |2 - y_3| < \frac{1}{2}.$$

So,

$$\frac{3}{2} < y_1 < \frac{5}{2}, \quad \frac{3}{2} < y_2 < \frac{5}{2} \quad \text{and} \quad \frac{3}{2} < y_3 < \frac{5}{2}.$$

Capitalizing on these bounds, it is immediate that $y \in X_P$. So, $(2, 2, 2) \in \text{int}(X_P)$. Hence,

$$(2, 2, 2) \notin \text{bn}(X_P).$$

Moreover, $(2, 2, 2) \in X_P^\circ$, by Proposition 2.3. Furthermore, $(2, 2, 2) \notin \partial X_P$, by Exercise 2.1. We now prove that

$$(12, 4, 2) \in \partial X_P.$$

Observe that ∂X_P is the set

$$\{(x_1, x_2, x_3) \in X_P : x_2 = 4\} \cup \cdots \cup \{(x_1, x_2, x_3) \in X_P : x_3 = 0\}.$$

So, $(12, 4, 2) \in \partial X_P$ since $(12, 4, 2)$ is in

$$\{(x_1, x_2, x_3) \in X_P : x_2 = 4\}.$$

Another way to show that $(12, 4, 2) \in \partial X_P$ is by proving that

$$(12, 4, 2) \in \text{bn}(X_P)$$

and then use Exercise 2.1. Hence, we must prove that for every real number $\varepsilon > 0$ there is $y \in B_\varepsilon(12, 4, 2) \cap X_P$ and there is $y \in B_\varepsilon(12, 4, 2) \cap (\mathbb{R}^3 \setminus X_P)$. Let $\varepsilon > 0$ be a real number. Observe that

$$(12, 4, 2) \in B_\varepsilon(12, 4, 2) \cap X_P.$$

On the other hand, consider the vector

$$\left(12 + \frac{\varepsilon}{2}, 4 + \frac{\varepsilon}{2}, 2 + \frac{\varepsilon}{2}\right).$$

Then,

$$\left(12 + \frac{\varepsilon}{2}, 4 + \frac{\varepsilon}{2}, 2 + \frac{\varepsilon}{2}\right) \in B_\varepsilon(12, 4, 2)$$

since

$$
\begin{aligned}
\|(12, 4, 2) - (12 + \tfrac{\varepsilon}{2}, 4 + \tfrac{\varepsilon}{2}, 2 + \tfrac{\varepsilon}{2})\| &= \|(-\tfrac{\varepsilon}{2}, -\tfrac{\varepsilon}{2}, -\tfrac{\varepsilon}{2})\| \\
&= (\frac{\varepsilon^2}{4} + \frac{\varepsilon^2}{4} + \frac{\varepsilon^2}{4})^{\frac{1}{2}} \\
&= \frac{\sqrt{3}}{2}\varepsilon \\
&< \varepsilon.
\end{aligned}
$$

On the other hand, it is immediate that

$$\left(12 + \frac{\varepsilon}{2}, 4 + \frac{\varepsilon}{2}, 2 + \frac{\varepsilon}{2}\right) \notin X_P.$$

The next result is an important property of the set of optimizers of a canonical optimization problem.

Proposition 2.4
Let P be a canonical optimization problem. If $c \neq 0$ then

$$S_P \cap X_P^\circ = \emptyset.$$

Proof:
The result is established by contradiction. Take $s \in S_P \cap X_P^\circ$. Then, $As < b$ and $s > 0$. Consider the family of vectors

$$\{s^\varepsilon\}_{\varepsilon \in \mathbb{R}^+}, \text{ where } s^\varepsilon = s + \varepsilon c^\mathsf{T}.$$

Then,
$$cs^\varepsilon > cs$$
for any $\varepsilon \in \mathbb{R}^+$, because $c\varepsilon c^\mathsf{T} = \varepsilon cc^\mathsf{T} > 0$, since $c \neq 0$.

We now show that there is ε such that $s^\varepsilon \in X_P$. We concentrate first on choosing a set D such that $s^\varepsilon \geq 0$ for every $\varepsilon \in D$. Let

$$D = \{\varepsilon \in \mathbb{R}^+ : \varepsilon \leq -\frac{s_j}{c_j} \text{ for } j = 1, \ldots, n \text{ whenever } c_j < 0\}.$$

Consider two cases:

(1) $c_j \geq 0$ for every j. Then, for every j, $s_j^\varepsilon = s_j + \varepsilon c_j \geq 0$ since $s_j > 0$ and $\varepsilon c_j \geq 0$. So, $s^\varepsilon \geq 0$ for every $\varepsilon \in \mathbb{R}^+ = D$.

(2) $c_j < 0$ for some j. Observe that

$$D = \left]0, \min\left\{-\frac{s_j}{c_j} : c_j < 0\right\}\right].$$

Let $k \in \{1, \ldots, n\}$. When $c_k \geq 0$, then $s_k^\varepsilon \geq 0$, similarly to (1). Otherwise,

$$s_k^\varepsilon = s_k + \varepsilon c_k \geq s_k + \left(-\frac{s_k}{c_k}\right)c_k = 0.$$

We now identify a set E such that $As^\varepsilon \leq b$ for every $\varepsilon \in E$. Let

$$E = \{\varepsilon \in \mathbb{R}^+ : \varepsilon \leq \frac{b_i - (As)_i}{(Ac^\mathsf{T})_i} \text{ for } i = 1, \ldots, m \text{ whenever } (Ac^\mathsf{T})_i > 0\}.$$

Consider two cases:

(1) $(Ac^\mathsf{T})_i \leq 0$ for every $i = 1, \ldots, m$. Then, $(As)_i + \varepsilon(Ac^\mathsf{T})_i < b_i$ since $(As)_i < b_i$ and $\varepsilon(Ac^\mathsf{T})_i \leq 0$. Thus, $As^\varepsilon \leq b$.

(2) $(Ac^\mathsf{T})_i > 0$ for some i. Observe that

$$E = \left]0, \min\left\{\frac{b_i - (As)_i}{(Ac^\mathsf{T})_i} : (Ac^\mathsf{T})_i > 0\right\}\right].$$

Let $k \in \{1, \ldots, m\}$. When $(Ac^\mathsf{T})_k \leq 0$, then $(As^\varepsilon)_k \leq b_k$, similarly to (1). Otherwise,

$$(As^\varepsilon)_k = (As)_k + \varepsilon(Ac^\mathsf{T})_k \leq (As)_k + \frac{b_k - (As)_k}{(Ac^\mathsf{T})_k}(Ac^\mathsf{T})_k = b_k.$$

In any case,
$$D \cap E \neq \emptyset.$$

Pick up $\varepsilon \in D \cap E$. Then, s^ε is admissible and $cs^\varepsilon > cs$ contradicting the hypothesis that $s \in S_P$.

QED

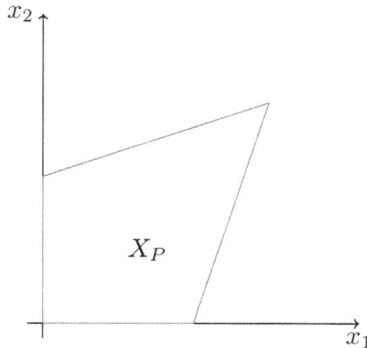

Figure 2.2: Boundary and interior of problem in Example 1.12

In other words, if $c \neq 0$ then the maximizers of the problem must lie on the boundary of the set of admissible vectors.

Example 2.3
Consider the canonical optimization problem P in Example 1.12 and the graphical representation of X_P in Figure 1.1. In Figure 2.2, we use different colors to represent the interior and the boundary of X_P. Since $c \neq 0$, by Proposition 2.4, any maximizer of P is in the darker part of the representation of X_P.

2.2   Existence

The objective of this section is to provide sufficient conditions for the existence of optimizers. We start by proving a result about the existence of optimizers of a canonical optimization problem when b is 0.

Proposition 2.5
In a canonical optimization problem, if $b = 0$ then either 0 is a maximizer or the objective map has no upper bound on the set of admissible vectors.

Proof:
Let P be a canonical optimization problem. Note that 0 is admissible. If 0 is a maximizer, the thesis follows immediately. Otherwise, observe that there is

$y \in X_P$ such that $cy > c0 = 0$. We want to show that

$$\{cx : x \in X_P\}$$

has no upper bound. Let $r \in \mathbb{R}$. Take z as the vector

$$\frac{|r| + 1}{cy} y.$$

Observe that $z \geq 0$, since $y \geq 0$ and $\frac{|r|+1}{cy} > 0$. Moreover,

$$Az = A\frac{|r| + 1}{cy} y = \frac{|r| + 1}{cy} Ay \leq 0.$$

So, $z \in X_P$ and $cz \in \{cx : x \in X_P\}$. On the other hand,

$$cz = c\frac{|r| + 1}{cy} y = \frac{|r| + 1}{cy} cy = |r| + 1 > r.$$

Thus, the objective map has no upper bound on X_P.   QED

As we discuss below, topological notions are useful in the study of the existence of optimizers. We start by proving that the set of admissible vectors and the set of optimizers of the canonical optimization problem are closed (see Definition 2.9).

Proposition 2.6
The set $\{x \in \mathbb{R}^n : x \geq 0\}$ is closed.

Proof:
Let $\{x^k\}_{k \in \mathbb{N}}$ be a sequence in $\{x \in \mathbb{R}^n : x \geq 0\}$ converging to $z \in \mathbb{R}^n$. Moreover, let $\{a^k\}_{k \in \mathbb{N}}$ be a sequence in $\mathbb{R}^n$ such that $a^k = 0$ for every $k \in \mathbb{N}$. Observe that $\{a^k\}_{k \in \mathbb{N}}$ converges to $0 \in \mathbb{R}^n$ by Proposition 2.17. On the other hand, $x^k \geq a^k$ for every $k \in \mathbb{N}$. Hence, by Proposition 2.16, $z \geq 0$. Thus, $z \in \{x \in \mathbb{R}^n : x \geq 0\}$. So, by Proposition 2.18, set $\{x \in \mathbb{R}^n : x \geq 0\}$ is closed.   QED

Proposition 2.7
Let A be an $m \times n$-matrix and $b \in \mathbb{R}^m$. Then, $\{x \in \mathbb{R}^n : Ax \leq b\}$ is closed.

Proof:
Let $\{x^k\}_{k \in \mathbb{N}}$ be a sequence in $\{x \in \mathbb{R}^n : Ax \leq b\}$ converging to $z \in \mathbb{R}^n$. Then,

$\{Ax^k\}_{k\in\mathbb{N}}$ is a sequence converging to $Az \in \mathbb{R}^m$ since $x \mapsto Ax : \mathbb{R}^n \to \mathbb{R}^m$ is continuous by Proposition 2.2. Moreover, let $\{a^k\}_{k\in\mathbb{N}}$ be a sequence in $\mathbb{R}^m$ such that $a^k = b$ for every $k \in \mathbb{N}$. Observe that $\{a^k\}_{k\in\mathbb{N}}$ converges to $b \in \mathbb{R}^m$, by Proposition 2.17. On the other hand, $Ax^k \le a^k$ for every $k \in \mathbb{N}$. Therefore, by Proposition 2.16, $Az \le b$. Hence, $z \in \{x \in \mathbb{R}^n : Ax \le b\}$. So, by Proposition 2.18, set $\{x \in \mathbb{R}^n : Ax \le b\}$ is closed.   QED

Proposition 2.8
The set of admissible vectors of a canonical optimization problem is closed.

Proof:
Let P be a canonical optimization problem. Observe that

$$X_P = \{x \in \mathbb{R}^n : x \ge 0\} \cap \{x \in \mathbb{R}^n : Ax \le b\}.$$

Therefore, X_P is closed by Proposition 2.19, since $\{x \in \mathbb{R}^n : x \ge 0\}$ is closed by Proposition 2.6 and $\{x \in \mathbb{R}^n : Ax \le b\}$ is closed by Proposition 2.7.   QED

Proposition 2.9
The set of optimizers of a canonical optimization problem is closed.

Proof:
Let P be a canonical optimization problem. If $S_P = \emptyset$ then S_P is closed (see Remark 2.2). Otherwise, let $\{s^k\}_{k\in\mathbb{N}}$ be a sequence in S_P converging to s. By Proposition 2.2, the map $x \mapsto cx$ is continuous and so the sequence $\{cs^k\}_{k\in\mathbb{N}}$ converges to cs. On the other hand, since each $s^k \in S_P$, then

$$cs^k = \max\{cx : x \in X_P\}$$

for every k. Hence, $\{cs^k\}_{k\in\mathbb{N}}$ is a constant sequence and so, by Proposition 2.15, $\{cs^k\}_{k\in\mathbb{N}}$ converges to $\max\{cx : x \in X_P\}$. Therefore,

$$cs = \max\{cx : x \in X_P\}.$$

On the other hand, $s \in X_P$, by Proposition 2.18, since, by Proposition 2.8, X_P is closed. Therefore, $s \in S_P$. So, by Proposition 2.18, S_P is closed.   QED

The following result, called the *Existence of Maximizer Theorem*, establishes a sufficient condition for the existence of a maximizer of a canonical optimization problem.

Theorem 2.1

When the set of admissible vectors is non-empty and bounded, there exists a maximizer for a canonical optimization problem.

Proof:

Let P be a canonical optimization problem. The set X_P is bounded, by hypothesis, and closed, by Proposition 2.8. Hence, X_P is compact (see Definition 2.12). Observe that, the objective map $x \mapsto cx$ is continuous, by Proposition 2.2. Thus, since $X_P \neq \emptyset$, by Proposition 2.28, the objective map $x \mapsto cx$ has a maximum in X_P. Therefore, P has a maximizer.   QED

Example 2.4

Recall the canonical optimization problem P in Example 2.2. We now show that X_P is bounded. We must prove that there is a positive real number μ such that

$$||x|| \leq \mu$$

for every $x \in X_P$. Take $\mu = 13$. Let $x \in X_P$. Then,

$$x_2 \leq 4, \quad x_3 \leq 3 \quad \text{and} \quad -2x_2 + x_1 \leq 4.$$

Hence,

$$x_1 \leq 4 + 2x_2 \leq 12.$$

Therefore,

$$||x|| = (x_1^2 + x_2^2 + x_3^2)^{\frac{1}{2}} \leq (12^2 + 4^2 + 3^2)^{\frac{1}{2}}.$$

On the other hand,

$$X_P \neq \emptyset$$

since $(2, 2, 2) \in X_P$, as shown in Example 2.2. Thus, by Theorem 2.1,

$$S_P \neq \emptyset.$$

Moreover, no maximizer of P is in X_P°, by Proposition 2.4 since $c \neq 0$ for P. Consider the problem P' obtained from P by removing the restriction $x_3 \leq 3$. We now show that $X_{P'}$ is not bounded. Suppose, by contradiction, that there is a positive real number μ such that

$$||x|| \leq \mu$$

for every $x \in X_{P'}$. Consider the vector $(1, 1, \mu)$. It is immediate that $(1, 1, \mu) \in X_{P'}$. On the other hand,

$$||(1, 1, \mu)|| = (1 + 1 + \mu^2)^{\frac{1}{2}} > \mu,$$

which is a contradiction. Therefore, $X_{P'}$ is not bounded. Hence, Theorem 2.1 cannot be applied to P'.

The next proposition uses Theorem 2.1 for standard optimization problems.

Proposition 2.10
Let P be a standard optimization problem with $X_P \neq \emptyset$. Then, $S_P \neq \emptyset$ provided that $X_{SC(P)}$ is a bounded set.

Proof:
Note that $X_{SC(P)} \neq 0$ since $X_{SC(P)} = X_P$ (see Proposition 1.2). Because, by hypothesis, $X_{SC(P)}$ is a bounded set, then, by Theorem 2.1, $S_{SC(P)} \neq \emptyset$. Thus, $S_P \neq \emptyset$, again by Proposition 1.2.   QED

2.3   Solved Problems and Exercises

Problem 2.1
Consider a canonical optimization problem $P = (A, b, c)$ such that X_P is bounded by

- line passing through points $(0, 1)$ and $(1, 2)$;

- line passing through points $(1, 2)$ and $(2, 1)$;

- line passing through points $(2, 1)$ and $(1, 0)$;

- x_1-axis and x_2-axis.

(1) Choose A and b in such a way that X_P contains the points $(1, 0)$ and $(0, 1)$.

(2) Show that X_P is bounded.

(3) Choose an interior point of X_P and show that it is in $\mathsf{int}(X_P)$.

Solution:
(1) Recall that the equation of a line has the form

$$x_2 = dx_1 + e.$$

So, the equation of

- the line passing through points $(0, 1)$ and $(1, 2)$ is $-x_1 + x_2 = 1$;

- the line passing through points $(1, 2)$ and $(2, 1)$ is $x_1 + x_2 = 3$;
- the line passing through points $(2, 1)$ and $(1, 0)$ is $x_1 - x_2 = 1$.

Hence, the restrictions for the set of admissible vectors contains:

- $-x_1 + x_2 \leq 1$ since the vector $(1, 0)$ is admissible;
- $x_1 + x_2 \leq 3$ since the vector $(1, 0)$ is admissible;
- $x_1 - x_2 \leq 1$ since the vector $(0, 1)$ is admissible.

Thus, A is the matrix

$$
\begin{bmatrix}
-1 & 1 \\
1 & 1 \\
1 & -1
\end{bmatrix}
$$

and b is the vector:

$$
b = \begin{bmatrix}
1 \\
3 \\
1
\end{bmatrix}
$$

(2) We must show that there is a positive real number μ such that $||x|| \leq \mu$ for every $x \in X_P$. Take $\mu = 5$ and let $x \in X_P$. Observe that

- $x_2 \leq 3$ because $x_1 + x_2 \leq 3$ and $x_1 \geq 0$;
- $x_1 \leq 3$ because $x_1 - x_2 \leq 1$ and $x_2 \geq 0$.

Therefore,

$$
||x|| \leq (3^2 + 3^2)^{\frac{1}{2}} \leq 5.
$$

(3) Consider the vector $(1, 1)$. It is immediate that this vector is in X_P°. Then, $(1, 1)$ is in $\text{int}(X)$ (see Proposition 2.3).   ◁

Exercise 2.2

Consider the following canonical optimization problem P

$$
\begin{cases}
\max\limits_{(x_1, x_2)} x_1 + 4x_2 \\
x_1 \leq 2 \\
2x_1 + x_2 \leq 7 \\
x_1, x_2 \geq 0.
\end{cases}
$$

(1) Show that the vector $(1, 1)$ is in $\text{int}(X_P)$.

(2) Investigate whether or not X_P is bounded.

(3) What can be concluded about the emptiness of S_P?

(4) Is there an optimizer of P in X_P°?

(5) Consider the problem P' obtained from P by removing the restriction $2x_1 + x_2 \leq 7$. Is $X_{P'}$ bounded? What can be said about the existence of optimizers of P'.

Exercise 2.3

Consider the following standard optimization problem P

$$\begin{cases} \min_{x} \; -2x_1 - 3x_2 \\ 2x_1 + x_2 + x_3 = 6 \\ x \geq 0. \end{cases}$$

Does P have a minimizer?

2.4   Relevant Background

In this section, we provide an overview of the topological notions and results over $\mathbb{R}^n$ relevant for the optimization material presented in this chapter. The interested reader can also consult [4, 48]. We start with the concept of norm.

Definition 2.2

A *norm* on the vector space $\mathbb{R}^n$ is a map

$$N : \mathbb{R}^n \to \mathbb{R}_0^+$$

such that

- $N(x) = 0$ if and only if $x = 0$;
- $N(\alpha x) = |\alpha| N(x)$, for any $\alpha \in \mathbb{R}$;
- $N(x + y) \leq N(x) + N(y)$.

The first requirement is called *positive definiteness*, the second *positive homogeneity*, and the last *triangular inequality*. The number

$$N(x)$$

is called the *norm* of vector x.

Definition 2.3
The *Euclidean norm* (on $\mathbb{R}^n$) is the map $\|\cdot\| : \mathbb{R}^n \to \mathbb{R}_0^+$ such that

$$\|x\| = \sqrt{\sum_{j=1}^{n} x_j^2}.$$

The Euclidean norm provides a distance from 0 to x, for every $x \in \mathbb{R}^n$.

Exercise 2.4
Show that $\|\cdot\| : \mathbb{R}^n \to \mathbb{R}_0^+$ when $n = 1$ coincides with the absolute value map $|\cdot| : \mathbb{R} \to \mathbb{R}_0^+$.

Definition 2.4
An *inner product* on the vector space $\mathbb{R}^n$ is a map

$$I : (\mathbb{R}^n)^2 \to \mathbb{R}$$

satisfying the following properties:

- $I(x, y) = I(y, x)$;

- $I(\alpha x, y) = \alpha I(x, y)$, for any $\alpha \in \mathbb{R}$;

- $I(x + y, z) = I(x, z) + I(y, z)$;

- $I(x, x) \geq 0$;

- $I(x, x) = 0$ if and only if $x = 0$.

The first requirement is called *symmetry*, the second and the third are called *linearity conditions*, and the fourth and the fifth are called *positive definiteness*. The usual inner product over $\mathbb{R}^n$ is as follows:

Definition 2.5
The *Euclidean inner product* (on $\mathbb{R}^n$) is the map $\cdot : (\mathbb{R}^n)^2 \to \mathbb{R}$ such that

$$x \cdot y = \sum_{j=1}^{n} x_j y_j.$$

Remark 2.1

Observe that

$$\|x\| = \sqrt{x \cdot x}.$$

Definition 2.6

The vectors $v^1, v^2 \in \mathbb{R}^n$ are *perpendicular* or *orthogonal*, indicated by

$$v^1 \perp v^2,$$

whenever $v^1 \cdot v^2 = 0$.

Example 2.5

The vectors $(1,1)$ and $(-1,1)$ are orthogonal.

Proposition 2.11 (Cauchy–Schwarz Inequality)

Given $x, y \in \mathbb{R}^n$, then

$$|x \cdot y| \le \|x\| \|y\|.$$

Proof:

Consider two cases:

(1) $y = 0$. Immediate, by inspection.

(2) $y \ne 0$. Then, $y \cdot y \ne 0$ by positive definiteness. Let z be the vector

$$x - \frac{x \cdot y}{y \cdot y} y.$$

Observe that

$$z \cdot y = x \cdot y - \frac{x \cdot y}{y \cdot y} y \cdot y = 0 \quad \text{and} \quad x = x - \frac{x \cdot y}{y \cdot y} y + \frac{x \cdot y}{y \cdot y} y = z + \frac{x \cdot y}{y \cdot y} y.$$

Therefore,

$$
\begin{aligned}
x \cdot x &= \left(z + \frac{x \cdot y}{y \cdot y} y\right) \cdot \left(z + \frac{x \cdot y}{y \cdot y} y\right) \\
&= z \cdot z + 2 \frac{x \cdot y}{y \cdot y} z \cdot y + \frac{(x \cdot y)^2}{(y \cdot y)^2} y \cdot y \\
&= z \cdot z + \frac{(x \cdot y)^2}{y \cdot y} \\
&\ge \frac{(x \cdot y)^2}{y \cdot y}
\end{aligned}
$$

and so the thesis follows immediately.   QED

The next results shows that the norm of a vector is an upper bound of the absolute value of each component of the vector.

Proposition 2.12
Let $x \in \mathbb{R}^n$. Then,
$$|x_j| \leq \|x\|$$
for every $j = 1, \ldots, n$.

Proof:
Then,
$$
\begin{aligned}
|x_j| &= |x \cdot e^j| \\
&\leq \|x\|\|e^j\| \quad \text{(by Proposition 2.11)} \\
&= \|x\|
\end{aligned}
$$
using the standard basis of $\mathbb{R}^n$ (see Definition 1.20). QED

Definition 2.7
The *open ball* of radius $\varepsilon > 0$ centered on $x \in \mathbb{R}^n$, denoted by
$$B_\varepsilon(x),$$
is the set $\{y \in \mathbb{R}^n : \|x - y\| < \varepsilon\}$. Moreover, the *closed ball* of radius ε centered on x, denoted by
$$B_\varepsilon[x],$$
is the set $\{y \in \mathbb{R}^n : \|x - y\| \leq \varepsilon\}$.

Example 2.6
Observe that
$$\left(\frac{1}{2}, \frac{1}{2}, \frac{1}{2}\right) \in B_1(1,1,1)$$
since
$$\|(1,1,1) - \left(\frac{1}{2}, \frac{1}{2}, \frac{1}{2}\right)\| = \|\left(\frac{1}{2}, \frac{1}{2}, \frac{1}{2}\right)\| = \sqrt{\frac{3}{4}} < 1.$$
On the other hand,
$$(0,0,0) \notin B_{\frac{1}{2}}(2,2,2).$$
Indeed:
$$\|(2,2,2) - (0,0,0)\| = \|(2,2,2)\| = \sqrt{12} > \frac{1}{2}.$$
Thus, $(0,0,0) \notin B_{\frac{1}{2}}(2,2,2)$.

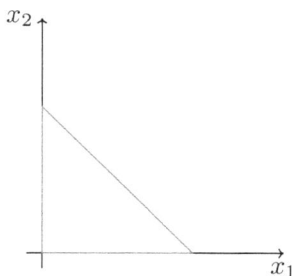

Figure 2.3: Boundary and interior of problem in Example 2.7

Definition 2.8
Let $x \in \mathbb{R}^n$ and $U \subseteq \mathbb{R}^n$. We say that x is an *interior point* of U if there is $\varepsilon > 0$ such that:
$$B_\varepsilon(x) \subset U.$$
Furthermore, we say that x is a *boundary point* of U if for every $\varepsilon > 0$ we have
$$B_\varepsilon(x) \cap U \neq \emptyset \ \text{ and } \ B_\varepsilon(x) \cap (\mathbb{R}^n \setminus U) \neq \emptyset.$$

Example 2.7
Let X be the set
$$\{(x_1, x_2) \in \mathbb{R}^2 : x_1 + x_2 \leq 2, x_1 \geq 0, x_2 \geq 0\}$$
depicted in Figure 2.3. Then,
$$\left(\frac{1}{2}, \frac{1}{2}\right) \text{ is an interior point}$$
and
$$(0, 1) \text{ is a boundary point.}$$

Notation 2.1
The set of all interior points of U is called the *interior* of U and is denoted by
$$\mathsf{int}(U).$$
Moreover, the set of all boundary points of U is called the *boundary* of U and is denoted by
$$\mathsf{bn}(U).$$

Definition 2.9
We say that a set $U \subseteq \mathbb{R}^n$ is *open* if $U = \mathsf{int}(U)$. The complement of an open set is called a *closed set*.

Remark 2.2
Observe that $\emptyset$ is an open and a closed set.

We now define convergence in $\mathbb{R}^n$.

Definition 2.10
A sequence $\{x^k\}_{k\in\mathbb{N}}$ in $\mathbb{R}^n$ *converges* to $x \in \mathbb{R}^n$ if for every real number $\delta > 0$ there is $k \in \mathbb{N}$ such that

$$\|x^m - x\| < \delta, \text{ for every } m > k.$$

Example 2.8
The sequence

$$\left\{ \left(-\frac{1}{k}, \frac{1}{k} \right) \right\}_{k\in\mathbb{N}}$$

converges to $(0,0)$.

Sometimes, the following characterization of convergence is also useful. For more details see [4].

Proposition 2.13
Let $\{x^k\}_{k\in\mathbb{N}}$ be a sequence in $\mathbb{R}^n$ and $x \in \mathbb{R}^n$. Then, $\{x^k\}_{k\in\mathbb{N}}$ converges to x if and only if, for each $j = 1, \ldots, n$, the sequence $\{(x^k)_j\}_{k\in\mathbb{N}}$ converges to x_j in $\mathbb{R}$.

That is, convergence in $\mathbb{R}^n$ amounts to pointwise convergence in $\mathbb{R}$.

Proposition 2.14
Let $\{u_k\}_{k\in\mathbb{N}}$ and $\{w_k\}_{k\in\mathbb{N}}$ be sequences in $\mathbb{R}$ such that there is $m \in \mathbb{N}$ with $u_n \leq w_n$ for every natural number $n > m$. Assume that $\{u_k\}_{k\in\mathbb{N}}$ and $\{w_k\}_{k\in\mathbb{N}}$ converge to u and w, respectively. Then, $u \leq w$.

Proposition 2.15
Let $\{u_k\}_{k\in\mathbb{N}}$ be a constant sequence in $\mathbb{R}$. Then, $\{u_k\}_{k\in\mathbb{N}}$ converges to u_0.

Taking account Proposition 2.13, counterparts of the previous results are stated for $\mathbb{R}^n$.

Proposition 2.16
Let $\{x^k\}_{k\in\mathbb{N}}$ and $\{y^k\}_{k\in\mathbb{N}}$ be sequences in $\mathbb{R}^n$ such that there is $m \in \mathbb{N}$ with $x^n \leq y^n$ for every natural number $n > m$. Assume that $\{x^k\}_{k\in\mathbb{N}}$ and $\{y^k\}_{k\in\mathbb{N}}$ converge to x and y, respectively. Then, $x \leq y$.

Proposition 2.17
Let $\{x^k\}_{k\in\mathbb{N}}$ be a constant sequence in $\mathbb{R}^n$. Then, $\{x^k\}_{k\in\mathbb{N}}$ converges to x^0.

The next result provides a useful characterization of closed sets.

Proposition 2.18
A set $U \subseteq \mathbb{R}^n$ is closed if and only if, for every sequence $\{x^k\}_{k\in\mathbb{N}}$ of elements of U, if $\{x^k\}_{k\in\mathbb{N}}$ converges to x in $\mathbb{R}^n$, then $x \in U$.

Observe that for showing that $x \in U$ when U is closed, it is enough to find a sequence in U that converges to x.

The following is an immediate consequence of Proposition 2.18.

Proposition 2.19
The class of closed sets is closed under finite union and intersection.

Proposition 2.20
A closed ball is a closed set.

Definition 2.11
We say that a set $U \subseteq \mathbb{R}^n$ is *bounded* if there exists $\mu \in \mathbb{R}^+$ such that $\|u\| \leq \mu$ for every $u \in U$.

Definition 2.12
A set $U \subseteq \mathbb{R}^n$ is *compact* if it is closed and bounded.

Example 2.9
For instance, the closed unit ball in $\mathbb{R}^n$ centered in 0 is compact; that is, the set

$$\{x \in \mathbb{R}^n : x_1^2 + \cdots + x_n^2 \leq 1\}$$

is compact.

Before proceeding we need some additional results on compact sets.

Proposition 2.21
Every sequence taking values on a compact set has a convergent subsequence in that set.

The following result provides a sufficient condition for a set to be compact.

Proposition 2.22
Let $U \subseteq \mathbb{R}^n$ be a set such that every infinite sequence in U has a subsequence converging to an element of U. Then, U is compact.

We now introduce the notion of continuous map which will be essential for some results in linear optimization.

Definition 2.13
A map $f : \mathbb{R}^n \to \mathbb{R}^m$ is *continuous at* x in $\mathbb{R}^n$ if, for every sequence $\{x^k\}_{k \in \mathbb{N}}$ in $\mathbb{R}^n$ that converges to x, $\{f(x^k)\}_{k \in \mathbb{N}}$ converges to $f(x)$. A map is *continuous* if it is continuous at every $x \in \mathbb{R}^n$.

The following results provide useful alternative characterizations of continuous maps.

Proposition 2.23
A map $f : \mathbb{R}^n \to \mathbb{R}^m$ is continuous at x if and only if, for every $\delta > 0$, there exists $\epsilon > 0$ such that if $\|y - x\| < \epsilon$ then $\|f(y) - f(x)\| < \delta$.

Proposition 2.24
A map $f : \mathbb{R}^n \to \mathbb{R}^m$ is continuous if and only if the map $f_i : \mathbb{R}^n \to \mathbb{R}$ such that $f_i(x) = (f(x))_i$ is continuous, for every $i = 1, \ldots, m$.

Proposition 2.25
The map $f = u \mapsto \|u - v\| : \mathbb{R}^n \to \mathbb{R}_0^+$ is continuous, where $v \in \mathbb{R}^n$.

The following result is also useful in the sequel.

Proposition 2.26
The inverse image of a closed set under a continuous map is closed.

The following result shows that compactness is preserved by continuous maps.

Proposition 2.27
Let $U \subseteq \mathbb{R}^n$ be a compact set and $f : \mathbb{R}^n \to \mathbb{R}$ a continuous map. Then, $f(U)$ is compact.

The following result provides a sufficient condition for a map to have a maximum. First we introduce the relevant notions.

Definition 2.14
A set $U \subseteq \mathbb{R}$ has an *upper bound* r whenever $u \leq r$ for every $u \in U$. A map $f : \mathbb{R}^n \to \mathbb{R}$ has an *upper bound* or is *bounded from above* whenever $\{f(x) : x \in \mathbb{R}^n\}$ has an upper bound. Furthermore, f has a *maximum* in $X \subseteq \mathbb{R}^n$ when there is $s \in X$ such that $f(x) \leq f(s)$ for every $x \in X$. Similarly for *lower bound*, *bounded from below* and *minimum*.

Proposition 2.28
Let $X \subseteq \mathbb{R}^n$ be a non-empty compact set and $f : \mathbb{R}^n \to \mathbb{R}$ a continuous map. Then, f has a maximum and has a minimum in X.

Chapter 3

Deciding Optimizers

The objective of this chapter is to provide a way for deciding whether or not a given admissible vector is an optimizer, relying on the Farkas' Lemma. Furthermore, we also use Farkas' Lemma to decide whether or not there are admissible vectors for the optimization problem at hand.

3.1 Farkas' Lemma

The purpose of this section is to prove the important result known as the Geometric Variant of the Farkas' Lemma as well as other variants relevant for optimization. Before that, we need to introduce the concepts of convex set, convex cone and primitive cone as well as some results about them.

Definition 3.1
A subset U of $\mathbb{R}^n$ is *convex* whenever U is closed under *convex combinations*; that is,

$$\alpha x + (1 - \alpha)y \in U$$

whenever $x, y \in U$ and $\alpha \in [0, 1]$.

Example 3.1
The vector $(\frac{4}{3}, 1)$ is a convex combination of vectors $(2, 3)$ and $(1, 0)$. Indeed:

$$\begin{bmatrix} \dfrac{4}{3} \\ 1 \end{bmatrix} = \frac{1}{3}\begin{bmatrix} 2 \\ 3 \end{bmatrix} + \frac{2}{3}\begin{bmatrix} 1 \\ 0 \end{bmatrix}.$$

Exercise 3.1

Show that $\emptyset$ is a convex set.

Definition 3.2

The *convex cone generated by* $\{u^1, \ldots, u^k\} \subset \mathbb{R}^n$,

$$\mathsf{C}(\{u^1, \ldots, u^k\}),$$

is the set

$$\left\{ \sum_{i=1}^{k} \alpha_i u^i : \alpha_i \in \mathbb{R}_0^+, 1 \leq i \leq k \right\}.$$

Exercise 3.2

Show that $\mathsf{C}(\{u^1, \ldots, u^k\})$ is a convex set.

Example 3.2

Let $\{e^1, \ldots, e^n\}$ be the canonical basis of $\mathbb{R}^n$ (recall Definition 1.20). Then,

$$\mathsf{C}(\{e^1, \ldots, e^n\}) = \{x \in \mathbb{R}^n : x \geq 0\}$$

as we now show.

($\subseteq$) Let $x \in \mathsf{C}(\{e^1, \ldots, e^n\})$. Then,

$$x = \sum_{j=1}^{n} \alpha_j e^j$$

for some $\alpha \in (\mathbb{R}_0^+)^n$. Hence, x_j is α_j for each $j = 1, \ldots, n$. Thus, $x \geq 0$.

($\supseteq$) Assume that $x \geq 0$. Observe that

$$x = \sum_{j=1}^{n} x_j e^j.$$

Moreover, $x_j \geq 0$ for every $j = 1, \ldots, n$. Hence, $x \in \mathsf{C}(\{e^1, \ldots, e^n\})$.

Definition 3.3

A subset C of $\mathbb{R}^n$ is a *convex cone* if there exists a finite subset U of $\mathbb{R}^n$ such that $C = \mathsf{C}(U)$.

Example 3.3
Recall Example 3.2. Then, the set

$$\{x \in \mathbb{R}^n : x \geq 0\}$$

is a convex cone.

Definition 3.4
A convex cone C is *primitive* if there exists a set U of linearly independent vectors such that $C = \mathsf{C}(U)$.

Example 3.4
Recall Example 3.3. Then, the convex cone

$$\{x \in \mathbb{R}^n : x \geq 0\}$$

is primitive.

Example 3.5
Observe that $\emptyset$ is a primitive convex cone since

$$\emptyset = \mathsf{C}(\emptyset)$$

and $\emptyset$ is a linearly independent set (see Example 1.26).

We now prove several results that are needed for Farkas' Lemma.

Proposition 3.1
Every primitive convex cone is closed.

Proof:
Let $C \subseteq \mathbb{R}^n$ be a primitive convex cone. Since C is a primitive cone, then $C = \mathsf{C}(U)$ for some set $U = \{u^1, \ldots, u^k\}$ of linearly independent vectors in $\mathbb{R}^n$. Let $h : \mathbb{R}^k \to \mathbb{R}^n$ be the map

$$h(x) = x_1 u^1 + \cdots + x_k u^k.$$

Let $\{e^1, \ldots, e^k\}$ be the canonical basis of $\mathbb{R}^k$ (recall Definition 1.20). We start by proving that

$$C = h(\mathsf{C}(\{e^1, \ldots, e^k\})).$$

($\subseteq$) Let $x \in C$. Then,

$$x = \alpha_1 u^1 + \cdots + \alpha_k u^k, \text{ for some } \alpha_1, \ldots, \alpha_k \in \mathbb{R}_0^+.$$

Observe that $\alpha \in C(\{e^1, \ldots, e^k\})$, by Example 3.2, where α is the vector in $\mathbb{R}^k$ with components $\alpha_1, \ldots, \alpha_k$. Hence,

$$x = h(\alpha)$$

and so $x \in h(C(\{e^1, \ldots, e^k\}))$.

($\supseteq$) Assume that $x \in h(C(\{e^1, \ldots, e^k\}))$. Then, by Example 3.2,

$$x = h(\alpha) \text{ for some } \alpha \geq 0, \alpha \in \mathbb{R}^k.$$

Therefore, $x \in C$.
We now show that

$$h \text{ is an injective map.}$$

Indeed, assume that $h(x) = h(y)$. Then,

$$(x_1 - y_1)u^1 + \cdots + (x_k - y_k)u^k = 0.$$

Hence, $(x_j - y_j) = 0$ for $j = 1, \ldots, k$ because U is a linearly independent set. Let $g : h(\mathbb{R}^k) \to \mathbb{R}^k$ be such that

$$g(y) \text{ is the unique } x \text{ such that } h(x) = y.$$

This map is well defined since $y = h(x)$ for some $x \in \mathbb{R}^k$ and there is only one such x, for each $y \in h(\mathbb{R}^k)$. Observe that g is a bijection. Moreover, $g^{-1} : \mathbb{R}^k \to h(\mathbb{R}^k)$ is such that

$$g^{-1}(x) = h(x)$$

for every $x \in \mathbb{R}^k$. Therefore,

$$C = g^{-1}(C(\{e^1, \ldots, e^k\})).$$

It is immediate that g is a linear map. Thus, by Proposition 2.1, g is continuous. Recall that $\{x \in \mathbb{R}^k : x \geq 0\}$ is a closed set by Proposition 2.6. Thus, $C(\{e^1, \ldots, e^k\})$ is closed by Proposition 3.2. Then, C is a closed set (see Proposition 2.26).   QED

The next result states that the set of primitive convex cones generates the class of convex cones.

Proposition 3.2

Every convex cone is a finite union of of primitive convex cones.

Proof:

Let C be a convex cone. Then,

$$C = \mathsf{C}(U)$$

for some finite set $U = \{u^1, \ldots, u^k\} \subset \mathbb{R}^n$. We will show that each element x of $\mathsf{C}(U)$ belongs to a primitive convex cone generated by a subset of U. The thesis follows from the fact that the number of primitive convex cones generated by subsets of U is finite (indeed there are 2^k subsets of U but not all of them may be linearly independent sets).

Take $x \in \mathsf{C}(U)$. Consider two cases:

(1) $x = 0$. Then, $x \in \mathsf{C}(\{u^1\})$. So, x belongs to a primitive convex cone generated by a subset of U.

(2) $x \neq 0$. Let $U_x = \{u^{i_1}, \ldots, u^{i_\ell}\} \subseteq U$ be such that $x \in \mathsf{C}(U_x)$ and

$$|U_x| = \min_V \{|V| : V \subseteq U, x \in \mathsf{C}(V)\}.$$

Then,

$$x = \alpha_1 u^{i_1} + \cdots + \alpha_\ell u^{i_\ell},$$

where $\alpha_j > 0$ for each $j = 1, \ldots, \ell$. Moreover, $U_x \neq \emptyset$. We now prove that

$$U_x \text{ is a linearly independent set.}$$

Assume, by contradiction, that U_x is not linearly independent. Hence, there exists $\beta \in \mathbb{R}^\ell$ such that $\beta_1 u^{i_1} + \cdots + \beta_\ell u^{i_\ell} = 0$ and $\beta_i \neq 0$ for some $i = 1, \ldots, \ell$. Pick j such that

$$\left|\frac{\alpha_j}{\beta_j}\right| = \min\left\{\left|\frac{\alpha_i}{\beta_i}\right| : \beta_i \neq 0 \,\wedge\, 1 \leq i \leq \ell\right\}.$$

Take

$$\gamma = \alpha - \frac{\alpha_j}{\beta_j}\beta.$$

Observe that there exists $i = 1, \ldots, k$ such that $\gamma_i = 0$. Indeed:

$$\gamma_j = \alpha_j - \frac{\alpha_j}{\beta_j}\beta_j = 0.$$

We now show that

$$\gamma_i = \alpha_i - \frac{\alpha_j}{\beta_j}\beta_i \geq 0, \text{ for every } i \neq j.$$

Consider the following cases:

(a) $\beta_i = 0$. Then,

$$\alpha_i - \frac{\alpha_j}{\beta_j}\beta_i = \alpha_i \geq 0.$$

(b) $\beta_j, \beta_i \in \mathbb{R}^-$. Then,

$$-\frac{\alpha_j}{\beta_j} \leq -\frac{\alpha_i}{\beta_i}.$$

Hence,

$$-\frac{\alpha_j}{\beta_j}\beta_i \geq -\frac{\alpha_i}{\beta_i}\beta_i,$$

and so

$$\gamma_i = \alpha_i - \frac{\alpha_j}{\beta_j}\beta_i \geq \alpha_i - \frac{\alpha_i}{\beta_i}\beta_i = 0.$$

(c) Either $\beta_j \in \mathbb{R}^-$ and $\beta_i \in \mathbb{R}^+$ or $\beta_j \in \mathbb{R}^+$ and $\beta_i \in \mathbb{R}^-$. Then,

$$\alpha_i - \frac{\alpha_j}{\beta_j}\beta_i \geq 0.$$

(d) $\beta_j, \beta_i \in \mathbb{R}^+$. Then,

$$\frac{\alpha_j}{\beta_j}\beta_i \leq \frac{\alpha_i}{\beta_i}\beta_i$$

and so

$$\gamma_i = \alpha_i - \frac{\alpha_j}{\beta_j}\beta_i \geq \alpha_i - \frac{\alpha_i}{\beta_i}\beta_i = 0.$$

Thus, $\gamma \geq 0$.

Therefore,

$$\gamma_1 u^{i_1} + \cdots + \gamma_\ell u^{i_\ell} =$$
$$(\alpha_1 - \frac{\alpha_j}{\beta_j}\beta_1)u^{i_1} + \cdots + (\alpha_\ell - \frac{\alpha_j}{\beta_j}\beta_\ell)u^{i_\ell} =$$
$$(\alpha_1 u^{i_1} + \cdots + \alpha_\ell u^{i_\ell}) - \frac{\alpha_j}{\beta_j}(\beta_1 u^{i_1} + \cdots + \beta_\ell u^{i_\ell}) = x,$$

thus contradicting the choice of U_x, because $V = \{u^{i_m} \in U_x : \gamma_m \neq 0\} \subsetneq U_x$ and $x \in \mathsf{C}(V)$.   QED

Proposition 3.3
Every convex cone is closed.

Proof:

Let C be a convex cone. Then, by Proposition 3.2, C is a finite union of primitive cones. Since, by Proposition 3.1, each primitive cone is a closed set and any finite union of closed sets is also a closed set (see Proposition 2.19), we conclude that C is a closed set. QED

The following result states an important fact concerning closed sets.

Proposition 3.4

Let $U \subseteq \mathbb{R}^n$ be a non-empty closed set and $v \in \mathbb{R}^n$. Then, there exists at least one element of U whose distance to v is minimal.

Proof:

Let $x \in U$ and
$$B = \{u \in U : \|u - v\| \leq \|x - v\|\}.$$

Observe that $x \in B$. We start by proving that B is a closed set. Observe that
$$B = B_{\|x-v\|}[v] \cap U,$$

where $B_{\|x-v\|}[v] = \{z \in \mathbb{R}^n : \|z - v\| \leq \|x - v\|\}$. Since U is closed by hypothesis and $B_{\|x-v\|}[v]$ is closed (see Proposition 2.20), then B is also closed because it is the intersection of closed sets (by Proposition 2.19). Moreover, B is a bounded set. Indeed, let $u \in B$. Then,

$$
\begin{aligned}
\|u\| &\leq \|u - v\| + \|v\| \quad &\text{triangular inequality} \\
&\leq \|x - v\| + \|v\| \quad &\text{definition of } B
\end{aligned}
$$

Therefore, B is a compact set (recall Definition 2.12). Consider the map
$$f = u \mapsto \|u - v\| : \mathbb{R}^n \to \mathbb{R}_0^+.$$

By Proposition 2.28, since f is continuous (see Proposition 2.25), it has a minimum x_v on the compact set B. Hence, x_v is the vector of B closest to v; that is,
$$\|x_v - v\| \leq \|u - v\|$$

for every $u \in B$. In particular, $\|x_v - v\| \leq \|x - v\|$ since $x \in B$. Finally, let $y \in U \setminus B$. Then,
$$\|y - v\| > \|x - v\| \geq \|x_v - v\|.$$

Thus, $\|x_v - v\| \leq \|u - v\|$ for every $u \in U$. QED

The next result is a direct consequence of the proposition above for convex cones.

Proposition 3.5

Let $U \subseteq \mathbb{R}^n$ be a non-empty convex cone and $v \in \mathbb{R}^n$. Then, there exists at least one element of U whose distance to v is minimal.

Proof:

Let U be a convex cone. Then, by Proposition 3.3, U is a closed set. Hence, by Proposition 3.4, we conclude the thesis.   QED

The next result called the *Geometric Variant of the Farkas' Lemma*, is particularly relevant for optimization. It was stated and proved by Julius Farkas (see [30]). Its generalization to topological vector spaces is known as the Hahn–Banach Theorem (see [57]).

Proposition 3.6

Let $U = \{u^1, \ldots, u^k\} \subset \mathbb{R}^n$ and $v \in \mathbb{R}^n$. Then, exactly one of the following alternatives holds:

- $v \in \mathsf{C}(U)$;

- there exists a non-null row vector $w \in \mathbb{R}^n$ such that $w u^i \geq 0$ for $i = 1, \ldots, k$ and $wv < 0$.

Proof:

Let z be an element of $\mathsf{C}(U)$ such that the distance from z to v is minimal (such a z exists by Proposition 3.5) and

$$w = (z - v)^{\mathsf{T}}.$$

We have two cases.

(1) w is the zero vector. Then, $v \in \mathsf{C}(U)$.

(2) w is not the zero vector. We start by showing that

$$wz = 0.$$

Suppose, by contradiction, that $wz \neq 0$. Then, $z \neq 0$. Consider the family

$$\{y^\alpha\}_{\alpha \in]0,1]}$$

when $wz > 0$, otherwise, consider the family

$$\{y^\alpha\}_{\alpha \in \mathbb{R}^-}$$

when $wz < 0$, where
$$y^\alpha = (1 - \alpha)z.$$

Then,
$$y^\alpha \in \mathsf{C}(U).$$

On the other hand,
$$
\begin{aligned}
y^\alpha - v &= (1 - \alpha)z - v \\
&= (1 - \alpha)z - (z - w^\mathsf{T}) \\
&= w^\mathsf{T} - \alpha z.
\end{aligned}
$$

Therefore,
$$
\begin{aligned}
\|y^\alpha - v\|^2 &= (w^\mathsf{T} - \alpha z)^\mathsf{T}(w^\mathsf{T} - \alpha z) \\
&= (w - \alpha z^\mathsf{T})(w^\mathsf{T} - \alpha z) \\
&= ww^\mathsf{T} - \alpha z^T w^T - \alpha wz + \alpha^2 z^T z \\
&= \|w\|^2 - 2\alpha wz + \alpha^2 \|z\|^2.
\end{aligned}
$$

Observe that $\alpha wz > 0$. Consider the set composed by the α's such that
$$2\alpha wz > \alpha^2 \|z\|^2\,;$$

that is,
$$\alpha < \frac{2wz}{\|z\|^2}.$$

Hence, if $wz > 0$ pick up
$$0 < \alpha < \min\left(1, \frac{2wz}{\|z\|^2}\right),$$

otherwise
$$\alpha < \frac{2wz}{\|z\|^2}.$$

Thus, in both cases,
$$\|y^\alpha - v\|^2 < \|w\|^2 = \|z - v\|^2$$

contradicting the choice of z as the vector at the minimal distance from v. Hence, $wz = 0$.

Then,

(a) $wv < 0$. Indeed
$$0 < ww^T = w(z - v) = wz - wv = -wv.$$

(b) $wx \geq 0$ for every $x \in C(U)$:

If $x = z$, then as proved before $wx = 0$. Hence, it remains to prove the thesis when $x \neq z$. Note that

$$
\begin{aligned}
-wx &= -wx + wz \\
&= -w(x - z) \\
&= (v - z)^\mathsf{T}(x - z).
\end{aligned}
$$

Therefore, we must show that $(v - z)^\mathsf{T}(x - z) \leq 0$. Assume, by contradiction, that $(v - z)^\mathsf{T}(x - z) > 0$. Consider the family

$$\{y^\alpha\}_{\alpha \in \mathbb{R}^+}, \text{ where } y^\alpha = \alpha x + (1 - \alpha)z.$$

Observe that $y^\alpha \in C(U)$ provided that $\alpha \leq 1$. On the other hand,

$$y^\alpha - v = \alpha x + w^\mathsf{T} - \alpha z = w^\mathsf{T} + \alpha(x - z).$$

Therefore,

$$
\begin{aligned}
\|y^\alpha - v\|^2 &= (y^\alpha - v)^\mathsf{T}(y^\alpha - v) \\
&= (w^\mathsf{T} + \alpha(x - z))^\mathsf{T}(w^\mathsf{T} + \alpha(x - z)) \\
&= \|w\|^2 + 2\alpha w(x - z) + \alpha^2\|x - z\|^2 \\
&= \|w\|^2 + 2\alpha(z - v)^\mathsf{T}(x - z) + \alpha^2\|x - z\|^2.
\end{aligned}
$$

Since $(z - v)^\mathsf{T}(x - z) < 0$, then

$$\|y^\alpha - v\|^2 < \|w\|^2 = \|z - v\|^2$$

provided that

$$-2\alpha(z - v)^\mathsf{T}(x - z) > \alpha^2\|x - z\|^2.$$

Choose α such that

$$0 < \alpha < \min\left(1, \frac{-2(z - v)^\mathsf{T}(x - z)}{\|x - z\|^2}\right).$$

Then, y^α contradicts the choice of z as minimizing distance to v.

It remains to prove that the two alternatives of the statement of the proposition do not hold at the same time. Assume, by contradiction that both alternatives hold. Then,

$$v = \alpha_1 u^1 + \cdots + \alpha_k u^k$$

for some $\alpha_1, \ldots, \alpha_k \in \mathbb{R}_0^+$. On the other hand, $wu^j \geq 0$ for every $j = 1, \ldots, k$. Hence,

$$wv = \alpha_1 wu^1 + \cdots + \alpha_k wu^k \geq 0$$

contradicting that $wv < 0$.   QED

In the sequel, we need the following notations.

Notation 3.1
Given a matrix A, we denote by

$$a_{\bullet j}$$

the j-th column of A and by

$$a_{i\bullet}$$

the i-th row of A.

Notation 3.2
Given an $m \times n$-matrix A, we denote by

$$\mathsf{C}(A)$$

the convex cone in $\mathbb{R}^n$ generated by the set $\{a_{1\bullet}, \ldots, a_{m\bullet}\}$ of lines of A.

We provide a version of *Farkas' Lemma for Matrices*.

Proposition 3.7
Let A be an $m \times n$-matrix and $v \in \mathbb{R}^n$. Then, exactly one of the following alternatives holds:

- $v \in \mathsf{C}(A)$;

- there is a non-null row vector $w \in \mathbb{R}^n$ with $wA^{\mathsf{T}} \geq 0^{\mathsf{T}}$ and $wv < 0$.

Proof:
Using Proposition 3.6 with $U = \{a_{1\bullet}, \ldots, a_{m\bullet}\}$, exactly one of the following alternatives holds:

- $v \in \mathsf{C}(U)$;

- there exists a non-null row vector $w \in \mathbb{R}^n$ such that $wa_{i\bullet}^{\mathsf{T}} \geq 0$ for $i = 1, \ldots, m$ and $wv < 0$.

The thesis follows since $\mathsf{C}(A) = \mathsf{C}(U)$ and $wa_{i\bullet}^{\mathsf{T}} \geq 0$ for $i = 1, \ldots, m$ is equivalent to saying that $wA^{\mathsf{T}} \geq 0^{\mathsf{T}}$. QED

Exercise 3.3

Let A be an $m \times n$-matrix and $v \in \mathbb{R}^m$. Show that exactly one of the following alternatives holds:

- $v \in \mathsf{C}(A^\mathsf{T})$;

- there exists a non-null row vector $w \in \mathbb{R}^m$ such that $wA \geq 0^\mathsf{T}$ and $wv < 0$.

3.2   Using Convex Cones

The objective of this section is to show that the Farkas' Lemma provides a test for deciding whether or not an admissible vector is a local maximum for the objective in a canonical optimization problem. Recall the pure form of a canonical optimization problem introduced in Notation 1.8.

Definition 3.5
Let P be a canonical problem in pure form and $x \in X_P$. The i-th line of $\underline{A}$ is *active in x* if
$$\underline{a}_{i\bullet}x = \underline{b}_i.$$

Notation 3.3
We denote by
$$\underline{A}^x$$
the matrix containing the lines of $\underline{A}$ that are active in x, where $x \in X_P$.

Example 3.6
Consider the canonical optimization problem P in Example 1.12 presented in pure form. The 1st and the 2nd lines of $\underline{A}$ are active in $(3,3) \in X_P$. Hence,
$$\underline{A}^{(3,3)} = \begin{bmatrix} 3 & -1 \\ -1 & 3 \end{bmatrix}$$
On the other hand, the 3rd and the 4th lines of $\underline{A}$ are active in $(0,0) \in X_P$. Hence,
$$\underline{A}^{(0,0)} = \begin{bmatrix} -1 & 0 \\ 0 & -1 \end{bmatrix}.$$

With the example above, the reader can start to infer that for any vector in the boundary of the set of admissible vectors there exists a line active in that vector. This is the case as we show now.

Proposition 3.8
Consider a canonical optimization problem P in pure form and $x \in X_P$. Then, $x \in \partial X_P$ if and only if there exists a line of $\underline{A}$ active in x.

Proof:

We just prove one of the implications. The other implication is proved in a similar way.

($\rightarrow$) Assume that $x \in \partial X_P$. There are two cases.

(1) There is $i \in \{1, \ldots .m\}$ such that $a_{i \bullet} x = b_i$. Then, the i-th line of $\underline{A}$ is active in x.

(2) There is $\ell \in \{m+1, \ldots, m+n\}$ such that $x_{\ell - m} = 0$. Thus, the ℓ-th line of $\underline{A}$ is active in x. QED

The aim is to provide a test for checking whether or not an admissible vector is a maximizer. For that, we introduce the concept of local maximum for maps from $\mathbb{R}^n$ into $\mathbb{R}$ as follows.

Definition 3.6

A vector x is a *local maximum* of $f : \mathbb{R}^n \to \mathbb{R}$ in $D \subseteq \mathbb{R}^n$ if there exists $\varepsilon > 0$ such that, for every $y \in D$,

$$\text{if } \|x - y\| \leq \varepsilon \text{ then } f(x) \geq f(y).$$

The following result, called the *Local Maximizer Theorem*, provides a way to decide whether or not a boundary vector is a local maximum, using convex cones.

Theorem 3.1

Let P be a canonical optimization problem in pure form. Then, for every $x \in \partial X_P$,

$$x \text{ is a local maximum of } x \mapsto cx \text{ in } X_P \quad \text{if and only if} \quad c^{\mathsf{T}} \in \mathsf{C}(\underline{A}^x).$$

Proof:

If $c = 0$ then the thesis follows immediately. Otherwise, we show the equivalent statement that exactly one of the following alternatives holds:

- $c^{\mathsf{T}} \in \mathsf{C}(\underline{A}^x)$ and x is a local maximum of the map $x \mapsto cx$ in X_P;

- $c^{\mathsf{T}} \notin \mathsf{C}(\underline{A}^x)$ and x is not a local maximum of the map $x \mapsto cx$ in X_P.

By Proposition 3.8, there is a line of $\underline{A}$ active in x. By applying Proposition 3.7 to matrix $\underline{A}^x$ and vector c^{T}, it follows that exactly one of the following alternatives holds:

(1) $c^\mathsf{T} \in \mathsf{C}(\underline{A}^x)$;

(2) there exists a non-null row vector $w \in \mathbb{R}^n$ such that $w(\underline{A}^x)^\mathsf{T} \geq 0^\mathsf{T}$ and $wc^\mathsf{T} < 0$.

In case (1), the thesis is established by contradiction as follows. Assume that x is not a local maximum of the map $x \mapsto cx$ in X_P. That is, for every $\varepsilon > 0$ there exists $y \in X_P$ such that $\|x - y\| \leq \varepsilon$ and $cx < cy$. Pick $\varepsilon > 0$ and y in these conditions.

Since $c^\mathsf{T} \in \mathsf{C}(\underline{A}^x)$, there exists $\alpha \geq 0$ such that $(\underline{A}^x)^\mathsf{T}\alpha = c^\mathsf{T}$; that is, $\alpha^\mathsf{T}\underline{A}^x = c$. Hence, from $cx < cy$ it follows that

$$\alpha^\mathsf{T}\underline{A}^x x < \alpha^\mathsf{T}\underline{A}^x y.$$

Thus, there exists i such that $\alpha_i > 0$ (otherwise $0 < 0$) and

$$(\underline{A}^x x)_i < (\underline{A}^x y)_i$$

as we now show. Assume, by contradiction, that for every j such that $\alpha_j > 0$ we have $(\underline{A}^x x)_j \geq (\underline{A}^x y)_j$. Then,

$$\alpha^\mathsf{T}\underline{A}^x x = \sum_{j \in \{j : \alpha_j > 0\}} \alpha_j(\underline{A}^x x)_j \geq \sum_{j \in \{j : \alpha_j > 0\}} \alpha_j(\underline{A}^x y)_j = \alpha^\mathsf{T}\underline{A}^x y,$$

contradicting $\alpha^\mathsf{T}\underline{A}^x x < \alpha^\mathsf{T}\underline{A}^x y$. Then, by definition of $\underline{A}^x$, for such an i,

$$b_i = \underline{a}_{i\bullet}x < \underline{a}_{i\bullet}y.$$

Therefore, it is not the case that $\underline{A}y \leq b$. Hence, $y \notin X_P$, contradicting the choice of y.

Assume now that $c^\mathsf{T} \notin \mathsf{C}(\underline{A}^x)$. Then, (2) holds. Take $\varepsilon > 0$. We show that it is possible to find $y \in \mathbb{R}^n$ such that:

(a) $\|x - y\| \leq \varepsilon$;

(b) $cx < cy$;

(c) $y \in X_P$.

Consider the family

$$\{y^\beta\}_{\beta \in \mathbb{R}^+}, \quad \text{where } y^\beta = x - \beta w^\mathsf{T}.$$

Observe that (a) is guaranteed by considering the subfamily

$$\{y^\beta\}_{\beta \in B}, \quad \text{where } B = \left\{\beta \in \mathbb{R}^+ : \beta \leq \frac{\varepsilon}{\|w^\mathsf{T}\|}\right\}.$$

Since $wc^{\mathsf{T}} < 0$ and $\beta > 0$, from

$$
\begin{aligned}
cy^\beta &= cx - \beta cw^{\mathsf{T}} \\
&= cx - \beta(wc^T)^{\mathsf{T}} \\
&> cx
\end{aligned}
$$

we conclude (b). Finally, to prove (c), that is, $\underline{A}y^\beta \leq \underline{b}$, it suffices to establish, for each i-th line of $\underline{A}$, that

$$\underline{a}_{i\bullet}y^\beta \leq \underline{b}_i.$$

Consider two cases:

(i) The i-th line of $\underline{A}$ is active in x:
In this case, $\underline{a}_{i\bullet}x = \underline{b}_i$. Then,

$$\underline{a}_{i\bullet}y^\beta = \underline{b}_i - \beta\underline{a}_{i\bullet}w^{\mathsf{T}}.$$

Since $w(\underline{A}^x)^{\mathsf{T}} \geq 0^{\mathsf{T}}$, in particular, $\underline{a}_{i\bullet}w_i^{\mathsf{T}} \geq 0$. Hence, because β is positive, $\underline{a}_{i\bullet}y^\beta \leq \underline{b}_i$.

(ii) The i-th line of $\underline{A}$ is not active in x:
In this case, $\underline{a}_{i\bullet}x < \underline{b}_i$. Then, we want to ensure that

$$\underline{a}_{i\bullet}y^\beta = \underline{a}_{i\bullet}x - \beta\underline{a}_{i\bullet}w^{\mathsf{T}} \leq \underline{b}_i.$$

If $\underline{a}_{i\bullet}w^{\mathsf{T}} \geq 0$ then $\underline{a}_{i\bullet}y^\beta \leq \underline{b}_i$. Otherwise $\underline{a}_{i\bullet}w^{\mathsf{T}} < 0$. In this case, consider the subfamily

$$\{y^\beta\}_{\beta \in B_i'}, \quad \text{where } B_i' = \left\{ \beta \in \mathbb{R}^+ : \beta \leq \frac{\underline{b}_i - \underline{a}_{i\bullet}x}{|\underline{a}_{i\bullet}w^{\mathsf{T}}|} \right\}.$$

Then, it is immediate that

$$\underline{a}_{i\bullet}y^\beta \leq \underline{b}_i.$$

By choosing y in the family

$$\{y^\beta\}_{\beta \in B \cap B'},$$

where

$$B' = \bigcap_{i \in I} B_i'$$

and I is the set

$$\{i \in \{1, \ldots, m+n\} : i\text{-th is not active in } x \text{ and } a_{i\bullet}w^{\mathsf{T}} < 0\}$$

we conclude that x is not a local maximum of $x \mapsto cx$ in X_P. QED

Example 3.7

Recall Example 3.6. Then, $(3,3)$ is a local maximum of $x \mapsto 2x_1 + x_2$ in X_P since $(2,1)^\mathsf{T} \in \mathsf{C}(\underline{A}^{(3,3)})$, by Theorem 3.1. Indeed,

$$\begin{bmatrix} 2 \\ 1 \end{bmatrix} = \alpha_1 \begin{bmatrix} 3 \\ -1 \end{bmatrix} + \alpha_2 \begin{bmatrix} -1 \\ 3 \end{bmatrix}$$

has a positive solution with $\alpha_1 = \dfrac{7}{8}$ and $\alpha_2 = \dfrac{5}{8}$. On the other hand, $(0,0)$ is not a local maximum of $x \mapsto 2x_1 + x_2$ in X_P since $(2,1)^\mathsf{T} \notin \mathsf{C}(\underline{A}^{(0,0)})$, by Theorem 3.1. Indeed,

$$\begin{bmatrix} 2 \\ 1 \end{bmatrix} = \alpha_1 \begin{bmatrix} -1 \\ 0 \end{bmatrix} + \alpha_2 \begin{bmatrix} 0 \\ -1 \end{bmatrix}$$

does not have a positive solution.

The following result is known as the *Maximizer Theorem*.

Theorem 3.2

Every local maximum of the objective map of a canonical optimization problem in the set of admissible vectors is a maximizer and vice versa.

Proof:

Let P be a canonical optimization problem. It is immediate that a maximizer is also a local maximum. The proof of the other implication is by contraposition. Assume that $x \in X_P$ is not a maximum of $x \mapsto cx$ in X_P. Then, there is $z \in X_P$ such that $cx < cz$. We must show that x is not a local maximum of $x \mapsto cx$ in X_P. That is, we must show that, for every $\varepsilon > 0$, there exists $y \in X_P$ such that $\|x - y\| \le \varepsilon$ and $cx < cy$. Let $\varepsilon > 0$. Consider the family

$$\{y^\beta\}_{\beta \in B}, \quad \text{where} \quad y^\beta = (1 - \beta)x + \beta z$$

with

$$B = \left\{ \beta \in \mathbb{R}^+ : \beta \le \frac{\varepsilon}{\|x - z\|} \right\}.$$

Each y^β satisfies $\|x - y^\beta\| \le \varepsilon$. Pick $\beta \in B$ such that $\beta < 1$. Then, y^β is a convex combination of elements of X_P. Moreover, $y^\beta \in X_P$ as we show now. Indeed, it is immediate that $y^\beta \ge 0$. Furthermore,

$$Ay^\beta = (1 - \beta)Ax + \beta Az \le (1 - \beta)b + \beta b = b.$$

Finally,
$$cy^\beta = (1 - \beta)cx + \beta cz > (1 - \beta)cx + \beta cx = cx$$
because $cz > cx$ and β is positive. QED

Example 3.8
Recall Example 3.7. Then, $(3,3)$ is a maximizer of P and $(0,0)$ is not a maximizer of P, taking into account Theorem 3.2.

We now provide a characterization, based on convex cones, for the set of admissible vectors of a standard optimization problem to be non-empty.

Proposition 3.9
Let P be a standard optimization problem. Then,
$$X_P \neq \emptyset \quad \text{if and only if} \quad b \in \mathsf{C}(A^{\mathsf{T}}).$$

Proof:
($\rightarrow$) Let A be an $m \times n$-matrix and $x \in X_P$. Then,
$$b_i = a_{i1}x_1 + \dots a_{in}x_n$$
for every $i = 1, \dots, m$ and $x_j \geq 0$ for every $j = 1, \dots, n$. We want to find $\alpha \in (\mathbb{R}_0^+)^n$ such that
$$b = \alpha_1 a_{\bullet 1} + \dots + \alpha_n a_{\bullet n}.$$
It is enough to take α_j as x_j for $j = 1, \dots, n$.
($\leftarrow$) This implication follows in a similar way. QED

The following results, for deciding the non-emptiness of the set of admissible vectors, are called the *Standard and the Canonical Variants of the Farkas' Lemma*, respectively.

Proposition 3.10
Let P be a standard optimization problem. Then,
$$X_P \neq \emptyset \quad \text{if and only if} \quad \text{for every } w \neq 0, wb \geq 0 \text{ when } wA \geq 0.$$

Proof:
Observe that, by Proposition 3.9,
$$X_P \neq \emptyset \quad \text{iff} \quad b \in \mathsf{C}(A^T).$$

By Exercise 3.3 and since the two possibilities are exclusive, we conclude that

$$b \in \mathsf{C}(A^T) \quad \text{iff} \quad \text{for every } w \neq 0, wb \geq 0 \text{ when } wA \geq 0.$$

Thus, the thesis follows.   QED

Proposition 3.11
Let P be a canonical optimization problem. Then,

$$X_P \neq \emptyset \quad \text{if and only if} \quad \text{for every } w \neq 0, wb \geq 0 \text{ when } wA \geq 0 \text{ and } w \geq 0.$$

Proof:
Using Proposition 1.3, we have that

$$X_P \neq \emptyset \quad \text{iff} \quad X_{CS(P)} \neq \emptyset.$$

On the other hand, Proposition 3.10 states that

$$X_{CS(P)} \neq \emptyset \quad \text{iff} \quad \text{for every } w \neq 0, wb \geq 0 \text{ when } w \left[\begin{array}{cc} A & I \end{array} \right] \geq 0.$$

Observe that
$$w \left[\begin{array}{cc} A & I \end{array} \right] \geq 0 \quad \text{iff} \quad wA \geq 0 \text{ and } w \geq 0$$
and so the thesis follows.   QED

Exercise 3.4

Prove the standard variant from the canonical variant for deciding the non-emptiness of the set of admissible vectors using the Farkas' Lemma.

3.3   Solved Problems and Exercises

Problem 3.1
Let $U \subseteq \mathbb{R}^n$ be a finite set and $\mathcal{V}$ the class of all $V \subseteq \mathbb{R}^n$ closed under sum and multiplication by non-negative scalars. Show that

$$\mathsf{C}(U) = \bigcap_{U \subseteq V, V \in \mathcal{V}} V.$$

Solution:
Let $U = \{u^1, \ldots, u^k\}$.

($\supseteq$) It is enough to show that $\mathsf{C}(U) \in \mathcal{V}$:

(1) $\mathsf{C}(U)$ is closed under sum of vectors. Let $x, y \in \mathsf{C}(U)$. Then,

$$x = \sum_{i=1}^{k} \alpha_i u^i \text{ and } y = \sum_{i=1}^{k} \beta_i u^i$$

for some $\alpha, \beta \in (\mathbb{R}_0^+)^n$. Therefore,

$$x + y = \sum_{i=1}^{k} (\alpha_i + \beta_i) u^i \in \mathsf{C}(U),$$

since $(\alpha_i + \beta_i) \in \mathbb{R}_0^+$ for $i = 1, \ldots, k$.

(2) $\mathsf{C}(U)$ is closed under multiplication by non-negative scalars. Let $x \in \mathsf{C}(U)$. Then,

$$x = \sum_{i=1}^{k} \alpha_i u^i$$

for some $\alpha \in (\mathbb{R}_0^+)^n$. Let $\beta \in \mathbb{R}_0^+$. Hence,

$$\beta x = \sum_{i=1}^{k} (\beta \alpha_i) u^i \in \mathsf{C}(U)$$

since $(\beta \alpha_i) \in \mathbb{R}_0^+$ for $i = 1, \ldots, k$.

($\subseteq$) We show that $\mathsf{C}(U) \subseteq V$ for any $V \in \mathcal{V}$ such that $U \subseteq V$. We start by observing that

$$\mathsf{C}(U) = \bigcup_{\ell=1}^{k} \bigcup_{\substack{U' \subseteq U \\ |U'| = \ell}} \mathsf{C}(U'),$$

We prove that

$$\mathsf{C}(U') \subseteq V$$

for every $V \in \mathcal{V}$ and U' such that $U' \subseteq U \subseteq V$ and $|U'| = \ell$, by induction on ℓ.

(Base) $\ell = 1$. Let $U' = \{u'\}$ be such that $U' \subseteq U$. Then, $\mathsf{C}(U')$ has the form

$$\{\alpha u' : \alpha \in \mathbb{R}_0^+\}$$

and so is contained in V since $U \subseteq V$ and V is closed under multiplication by non-negative scalars.

(Step) Assume that $\mathsf{C}(U'') \subseteq V$ for every U'' such that $U'' \subseteq U$ and $|U''| \leq \ell$. Let
$$U' = \{u'^{\,1}, \ldots, u'^{\,\ell+1}\}$$
be such that $U' \subseteq U$. Take $x \in \mathsf{C}(U')$. Then, there is $\alpha \in (\mathbb{R}_0^+)^{\ell+1}$ such that

$$x = \sum_{i=1}^{\ell+1} \alpha_i u'^{\,i} = \left(\sum_{i=1}^{\ell} \alpha_i u'^{\,i} \right) + \alpha_{\ell+1} u'^{\,\ell+1}.$$

Then,

- $\displaystyle\sum_{i=1}^{\ell} \alpha_i u'^{\,i} \in V$, using the induction hypothesis;

- $\alpha_{\ell+1} u'^{\,\ell+1} \in V$, since $u'^{\,\ell+1} \in V$ and V is closed under multiplication by non-negative scalars.

Hence, $x \in V$, since V is closed under sum of vectors.   ◁

Problem 3.2

Consider the following canonical optimization problem P:

$$\begin{cases} \max_{x} \; 2x_1 + 3x_2 \\ x_1 + x_2 \leq 3 \\ -x_1 + x_2 \leq -1 \\ x \geq 0. \end{cases}$$

(1) Present P in pure form.

(2) Given $v \in \mathbb{R}^2$, discuss the different possibilities for $\underline{A}^v$.

(3) Find the lines of $\underline{A}$ active in $(2, 0)$ and the lines of $\underline{A}$ active in $(2, 1)$ and conclude whether or not these vectors are maximizers.

Solution:

(1) The pure form of P is:

$$\begin{cases} \max_{x} \begin{bmatrix} 2 & 3 \end{bmatrix} x \\ \begin{bmatrix} 1 & 1 \\ -1 & 1 \\ -1 & 0 \\ 0 & -1 \end{bmatrix} x \leq \begin{bmatrix} 3 \\ -1 \\ 0 \\ 0 \end{bmatrix}. \end{cases}$$

(2) Let $x = (x_1, x_2) \in X_P$. There are the following possibilities for $\underline{A}^x$:

- $\underline{A}^x$ is the empty matrix. Then, there are no active lines for x and so, by Proposition 3.8, $x \in X_P^\circ$;

- $\underline{A}^x$ is the matrix $\begin{bmatrix} 0 & -1 \end{bmatrix}$. Then, by Proposition 3.8, $x \in \partial X_P$ and is in the x_1-axis;

- $\underline{A}^x$ is the matrix $\begin{bmatrix} 1 & 1 \end{bmatrix}$. Then, by Proposition 3.8, $x \in \partial X_P$. Moreover, x is in the segment defined by $(2, 1)$ and $(3, 0)$;

- $\underline{A}^x$ is the matrix $\begin{bmatrix} -1 & 1 \end{bmatrix}$. Then, by Proposition 3.8, $x \in \partial X_P$. Moreover, x is in the segment defined by $(1, 0)$ and $(2, 1)$;

- $\underline{A}^x$ is the matrix $\begin{bmatrix} 1 & 1 \\ -1 & 1 \end{bmatrix}$. Then, by Proposition 3.8, $x \in \partial X_P$ and x is $(2, 1)$;

- $\underline{A}^x$ is the matrix $\begin{bmatrix} 1 & 1 \\ 0 & -1 \end{bmatrix}$. Then, by Proposition 3.8, $x \in \partial X_P$ and x is $(3, 0)$;

- $\underline{A}^x$ is the matrix $\begin{bmatrix} -1 & 1 \\ 0 & -1 \end{bmatrix}$. Then, by Proposition 3.8, $x \in \partial X_P$ and x is $(1, 0)$.

Observe that

$$\begin{bmatrix} -1 & 0 \end{bmatrix}$$

cannot be a line of $\underline{A}^x$ since X_P does not intersect the x_2-axis.

(3) The 4-th line is the unique active in $(2, 0)$. On the other hand, the 1st and the 2nd lines are active in $(2, 1)$. Therefore,

$$\underline{A}^{(2,0)} = \begin{bmatrix} 0 & -1 \end{bmatrix}$$

and

$$\underline{A}^{(2,1)} = \begin{bmatrix} 1 & 1 \\ -1 & 1 \end{bmatrix}.$$

Using Theorem 3.2, to prove whether or not $x \in X_P$ is a maximizer, we must show that x is a local maximum for the objective map in X_P. Moreover, by Theorem 3.1 it is enough to verify whether or not $c^\mathsf{T} \in \mathsf{C}(\underline{A}^x)$.

- $(2, 0)$: in this case, there is no $\alpha \in \mathbb{R}_0^+$ such that

$$\begin{bmatrix} 2 \\ 3 \end{bmatrix} = \alpha \begin{bmatrix} 0 \\ -1 \end{bmatrix}.$$

Hence, $(2, 0)$ is not a maximizer.

- $(2,1)$: in this case, we show that there are α_1 and α_2 in $\mathbb{R}_0^+$ such that

$$\begin{bmatrix} 2 \\ 3 \end{bmatrix} = \alpha_1 \begin{bmatrix} 1 \\ 1 \end{bmatrix} + \alpha_2 \begin{bmatrix} -1 \\ 1 \end{bmatrix}.$$

Indeed, take

$$\begin{cases} \alpha_1 = \frac{5}{2} \\ \alpha_2 = \frac{1}{2}. \end{cases}$$

Thus, $(2,1)$ is a maximizer.   ◁

Problem 3.3
Let P be the following canonical optimization problem

$$\begin{cases} \max_x\ 2x_1 + 5x_2 \\ -\frac{4}{3}x_1 - x_2 \leq -2 \\ 2x_1 + x_2 \leq 10 \\ x \geq 0. \end{cases}$$

Show that X_P is non-empty using convex cones.

Solution:
Observe that $CS(P)$ is as follows:

$$\begin{cases} \min_x\ -2x_1 - 5x_2 \\ -\frac{4}{3}x_1 - x_2 + x_3 = -2 \\ 2x_1 + x_2 + x_4 = 10 \\ x \geq 0. \end{cases}$$

Hence, $A_{CS(P)}$ and $b_{CS(P)}$ are:

$$\begin{bmatrix} -\frac{4}{3} & -1 & 1 & 0 \\ 2 & 1 & 0 & 1 \end{bmatrix} \quad \text{and} \quad \begin{bmatrix} -2 \\ 10 \end{bmatrix},$$

respectively. Using Proposition 3.9, for proving that $X_{CS(P)} \neq \emptyset$ it is enough to show that

$$b_{CS(P)} \in \mathsf{C}(A_{CS(P)}^{\mathsf{T}}).$$

Therefore, we must find $\alpha_1, \alpha_2, \alpha_3, \alpha_4 \in \mathbb{R}_0^+$ such that

$$\alpha_1 \begin{bmatrix} -\frac{4}{3} \\ 2 \end{bmatrix} + \alpha_2 \begin{bmatrix} -1 \\ 1 \end{bmatrix} + \alpha_3 \begin{bmatrix} 1 \\ 0 \end{bmatrix} + \alpha_4 \begin{bmatrix} 0 \\ 1 \end{bmatrix} = \begin{bmatrix} -2 \\ 10 \end{bmatrix}.$$

That is, we must find a non-negative solution for the following system of equations:

$$\begin{cases} -\dfrac{4}{3}\alpha_1 - \alpha_2 + \alpha_3 = -2 \\ 2\alpha_1 + \alpha_2 + \alpha_4 = 10. \end{cases}$$

For instance, take

$$\alpha_1 = 0, \ \alpha_2 = 2, \ \alpha_3 = 0 \text{ and } \alpha_4 = 8.$$

Thus, by Proposition 3.9, $X_{CS(P)} \neq \emptyset$. Then,

$$X_P \neq \emptyset$$

by Proposition 1.3. ◁

Exercise 3.5

Let A be an $m \times n$-matrix and $x \in \mathbb{R}^n$. Show that exactly one of the following alternatives holds:

- $x \in C(A)$;
- there exists a non-null row vector $e \in \mathbb{R}^n$ such that $eA^\mathsf{T} \geq 0^\mathsf{T}$ and $ex < 0$.

Exercise 3.6

Consider the following canonical optimization problem P:

$$\begin{cases} \max\limits_{x} \ 2x_1 + x_2 \\ -x_1 - x_2 \leq -1 \\ x_1 + x_2 \leq 3 \\ x_1 - x_2 \leq 1 \\ -x_1 + x_2 \leq 1 \\ -x_1 \leq 0 \\ -x_2 \leq 0. \end{cases}$$

Decide whether or not $(2,1)$ and $\left(\dfrac{3}{2}, \dfrac{1}{2}\right)$ are maximizers of P.

Exercise 3.7

Consider the following canonical optimization problem:

$$\begin{cases} \max_{x} 2x_1 + x_2 \\ -x_1 + x_2 \leq 1 \\ x_1 + x_2 \leq 3 \\ x_1 - x_2 \leq 1 \\ x \geq 0. \end{cases}$$

(1) Verify that $(2,1)$ is a local maximum and that $(1,2)$ is not a local maximum.

(2) Find an objective map for which $(1,2)$ is a local maximum.

(3) Is there an objective map such that all points in an edge are local maxima? Justify your answer.

Chapter 4

Computing Optimizers

The main objective of this chapter is to obtain a constructive counterpart of Theorem 2.1 by providing the means to calculate optimizers, using techniques from linear algebra. The reader can find, in Section 4.5, a brief survey of the relevant background including concepts and results. Throughout this chapter, we focus on standard optimization problems.

4.1   Basic Vectors

The main objective of this section is to introduce the important subset of the set of admissible vectors composed by the so called basic vectors. For that, it is useful to consider restricted standard optimization problems. We will show that for every optimization problem there is always an equivalent restricted standard problem. Recall that each line of matrix A is non-null (see Remark 1.2).

Definition 4.1
A standard optimization problem

$$
\begin{cases}
\min_{x} cx \\
Ax = b \\
x \geq 0,
\end{cases}
$$

where A is an $m \times n$-matrix, is *restricted* whenever

1. $m < n$;

2. $\mathrm{rank}(A) = m$.

We start by showing that the standard optimization problem resulting from applying the map CS (see Proposition 1.3) to a canonical optimization problem is restricted.

Proposition 4.1

Let P be a canonical optimization problem. Then,

$$CS(P)$$

is a restricted standard optimization problem.

Proof:

Assume that P is

$$\begin{cases} \max_{x} \; cx \\ Ax \leq b \\ x \geq 0. \end{cases}$$

Then, $CS(P)$ is

$$\begin{cases} \min_{y} \begin{bmatrix} -c & 0 \end{bmatrix} y \\[2mm] \begin{bmatrix} A & I \end{bmatrix} y = \begin{bmatrix} b \end{bmatrix} \\[2mm] y \geq 0, \end{cases}$$

where I is the $m \times m$ identity matrix. Note that the dimension of the matrix of $CS(P)$ is $m \times (n{+}m)$. Therefore, $m < n + m$. Moreover, the rank of the matrix is m because it includes the identity matrix of dimension m.   QED

The following result shows that every standard optimization problem has an equivalent restricted counterpart (recall the definition of map SC in Proposition 1.2).

Proposition 4.2

Let P be a standard optimization problem. Then,

$$CS(SC(P))$$

is a restricted standard optimization problem.

Proof:

It is enough to use Proposition 4.1 over $SC(P)$.   QED

Remark 4.1
From now on, we assume, without loss of generality, that any standard optimization problem is restricted, except when otherwise stated.

Notation 4.1
Given an $m \times n$-matrix, we denote by

$$M = \{1, \ldots, m\} \quad \text{and} \quad N = \{1, \ldots, n\}$$

the sets of row and column indices, respectively.

Notation 4.2
Given a matrix A and $B \subseteq N$, we denote by

$$A_B$$

the submatrix (see Definition 1.23) of A where the columns are the ones with indices in B.

Definition 4.2
Let A be an $m \times n$ matrix of a standard optimization problem. An *(index)* *basis* for A is a subset B of N with cardinality m such that the set of columns in A_B is linearly independent (that is, A_B is nonsingular).

Exercise 4.1

Show that

$$\binom{n}{m} = \frac{n!}{m!(n-m)!}$$

is an upper bound on the number of basis.

Example 4.1
Consider the following standard optimization problem

$$\begin{cases} \min_x \begin{bmatrix} -1 & -4 & 0 & 0 \end{bmatrix} x \\ \\ \begin{bmatrix} -1 & 1 & 1 & 0 \\ 2 & 1 & 0 & 1 \end{bmatrix} x = \begin{bmatrix} 1 \\ 6 \end{bmatrix} \\ \\ x \geq 0. \end{cases}$$

Observe that $N = \{1, 2, 3, 4\}$. The subsets of N that are candidates to basis are the following:

$$\{1, 2\}, \{1, 3\}, \{1, 4\}, \{2, 3\}, \{2, 4\}, \{3, 4\}$$

corresponding to the matrices

$$A_{\{1,2\}} = \begin{bmatrix} -1 & 1 \\ 2 & 1 \end{bmatrix}, \quad A_{\{1,3\}} = \begin{bmatrix} -1 & 1 \\ 2 & 0 \end{bmatrix}, \quad A_{\{1,4\}} = \begin{bmatrix} -1 & 0 \\ 2 & 1 \end{bmatrix},$$

$$A_{\{2,3\}} = \begin{bmatrix} 1 & 1 \\ 1 & 0 \end{bmatrix}, \quad A_{\{2,4\}} = \begin{bmatrix} 1 & 0 \\ 1 & 1 \end{bmatrix}, \quad A_{\{3,4\}} = \begin{bmatrix} 1 & 0 \\ 0 & 1 \end{bmatrix}.$$

It is immediate that the determinant of each matrix is non-null. Hence, each candidate is a basis.

This terminology is justified by the following result:

Proposition 4.3
Consider a standard optimization problem. Let $B \subseteq N$ be an index basis for A. Then, the columns in A_B constitute a basis for $\text{span}(A)$.

Proof:
Since B is an index basis for A, then the set composed by the columns of A_B is linearly independent. Moreover, $\text{rank}(A) = \text{rank}(A_B)$. Observe that, by Proposition 4.24, the dimension of $\text{span}(A)$ is $\text{rank}(A)$. So, the set composed by the columns of A_B is a basis for $\text{span}(A)$, by Proposition 4.12.   QED

Among the admissible vectors of a standard optimization problem, some are of particular significance.

Definition 4.3
Consider a standard optimization problem. An admissible vector x is *basic* whenever there exists $B \subseteq N$ such that:

- B is an (index) basis;

- $x_j = 0$ for every $j \in (N \setminus B)$.

In this case, B is *admissible*. Moreover, x is *degenerate* when the number of zero components of x is strictly greater than $|N \setminus B|$.

Example 4.2
Recall the standard optimization problem P introduced in Example 4.1. We now investigate which bases are admissible.

(1) $B = \{1, 2\}$. Let $x \in \mathbb{R}^4$ be such that $x_3 = 0$ and $x_4 = 0$. Observe that

$$\begin{bmatrix} -1 & 1 & 1 & 0 \\ 2 & 1 & 0 & 1 \end{bmatrix} \begin{bmatrix} x_1 \\ x_2 \\ 0 \\ 0 \end{bmatrix} = \begin{bmatrix} 1 \\ 6 \end{bmatrix}$$

if and only if

$$\begin{cases} -x_1 + x_2 = 1 \\ 2x_1 + x_2 = 6 \end{cases} \quad \text{if and only if} \quad \begin{cases} x_2 = \dfrac{8}{3} \\ x_1 = \dfrac{5}{3}. \end{cases}$$

Hence, $x \in X_P$. So, this basis is admissible with basic admissible vector

$$\left(\frac{5}{3}, \frac{8}{3}, 0, 0 \right).$$

This vector is non-degenerate.

(2) $B = \{1, 3\}$. Let $x \in \mathbb{R}^4$ be such that $x_2 = 0$ and $x_4 = 0$. Observe that

$$\begin{bmatrix} -1 & 1 & 1 & 0 \\ 2 & 1 & 0 & 1 \end{bmatrix} \begin{bmatrix} x_1 \\ 0 \\ x_3 \\ 0 \end{bmatrix} = \begin{bmatrix} 1 \\ 6 \end{bmatrix}$$

if and only if

$$\begin{cases} -x_1 + x_3 = 1 \\ 2x_1 = 6 \end{cases} \quad \text{if and only if} \quad \begin{cases} x_3 = 4 \\ x_1 = 3. \end{cases}$$

Hence, $x \in X_P$. Therefore, this basis is admissible with basic admissible vector

$$(3, 0, 4, 0).$$

This vector is non-degenerate.

(3) $B = \{1, 4\}$. Let $x \in \mathbb{R}^4$ be such that $x_2 = 0$ and $x_3 = 0$. Note that

$$
\begin{bmatrix} -1 & 1 & 1 & 0 \\ 2 & 1 & 0 & 1 \end{bmatrix}
\begin{bmatrix} x_1 \\ 0 \\ 0 \\ x_4 \end{bmatrix}
= \begin{bmatrix} 1 \\ 6 \end{bmatrix}
$$

if and only if

$$
\begin{cases} -x_1 &= 1 \\ 2x_1 + x_4 &= 6. \end{cases}
$$

So, $x \notin X_P$. Thus, the basis $B = \{1, 4\}$ is not admissible.

We omit details for the other bases. Anyhow, the admissible bases are:

$$\{1, 2\}, \quad \{1, 3\}, \quad \{2, 4\} \quad \text{and} \quad \{3, 4\},$$

with the following basic admissible vectors:

$$\left(\frac{5}{3}, \frac{8}{3}, 0, 0 \right), \quad (3, 0, 4, 0), \quad (0, 1, 0, 5) \quad \text{and} \quad (0, 0, 1, 6),$$

respectively.

Proposition 4.4
There is at most one basic admissible vector for each basis of a standard optimization problem.

Proof:
For $x \in \mathbb{R}^n$ to be admissible, it is necessary that $Ax = b$; that is, $A_B x_B + A_{N \backslash B} x_{N \backslash B} = b$. Then, for B to be a basis for x, it is necessary that $A_B x_B + A_{N \backslash B} 0 = b$; that is, $A_B x_B = b$. This system has exactly one solution, since A_B is nonsingular. If this solution (in $\mathbb{R}^m$) is non-negative, then adding zeros in the adequate positions yields an admissible vector basic for B.   QED

Proposition 4.5
The set of basic admissible vectors of a standard optimization problem is finite.

Proof:
The thesis follows from Exercise 4.1 and Proposition 4.4.   QED

Notation 4.3

Given $x \in \mathbb{R}^n$, we denote by

$$P_x$$

the set $\{j \in N : x_j > 0\}$.

The following results provide characterizations of basic admissible vectors.

Proposition 4.6

Consider a standard optimization problem with matrix A and let x be an admissible vector. Then, x is basic if and only if the set of columns of A with indices in P_x is linearly independent.

Proof:

($\rightarrow$) Assume that x is basic and $B \subseteq N$ is a basis for x. Thus, $x_j = 0$ for each $j \in (N \setminus B)$. Hence, $P_x \subseteq B$. Therefore, since the set $\{a_{\bullet j} : j \in B\}$ is linearly independent, so is the set $\{a_{\bullet j} : j \in P_x\}$.

($\leftarrow$) Assume that the set of columns with indices in P_x (that is, the set $\{a_{\bullet j} : j \in P_x\}$) is linearly independent. We consider two cases:

(1) $|P_x| = \operatorname{rank}(A)$. Then, P_x is an index basis for x. Hence, x is basic.

(2) $|P_x| < \operatorname{rank}(A)$. By Proposition 4.11, there is a set of columns of A constituting a basis for $\operatorname{span}(A)$ and containing $\{a_{\bullet j} : j \in P_x\}$. Let B be the set of indices of the columns in that set. Then, $B \supset P_x$ and $|B| = \operatorname{rank}(A)$. Finally, $x_j = 0$ for each $j \in (N \setminus B)$ follows from $(N \setminus B) \subset (N \setminus P_x)$.   QED

Proposition 4.7

Consider a standard optimization problem and let x be an admissible vector. Then, x is basic if and only if for every admissible vectors y and z, whenever there exists $\alpha \in]0, 1[$ such that $x = \alpha y + (1 - \alpha)z$ then $x = y = z$.

Proof:

Let A be the matrix of the problem.

($\rightarrow$) Assume that x is basic. Let y and z be admissible vectors and $\alpha \in]0, 1[$ such that

$$x = \alpha y + (1 - \alpha)z.$$

Let B be an admissible basis for x. Hence, $x_j = 0$ for every $j \in N \setminus B$. Thus,

$$y_j = z_j = 0$$

for every $j \in N \setminus B$, since $x_j = \alpha y_j + (1 - \alpha)z_j$, $y_j \geq 0$, $z_j \geq 0$, $\alpha y_j \geq 0$, $(1 - \alpha) \geq 0$ and $(1 - \alpha)z_j \geq 0$. Therefore,

$$A_B x_B = A_B y_B = A_B z_B = b$$

since $Ax = Ay = Az = b$. So,

$$A_B(x_B - y_B) = A_B(x_B - z_B) = 0.$$

Because A_B is nonsingular there is a unique w such that $A_B w = 0$. Hence,

$$x_B - y_B = x_B - z_B$$

and so $y_B = z_B$. Thus, $y = z$ and because $x = \alpha y + (1 - \alpha)z$ we have $x = y = z$.

($\leftarrow$) Assume that, for every admissible vectors y and z if

$$x = \alpha y + (1 - \alpha)z$$

for some $\alpha \in \,]0,1[$ then $x = y = z$. Suppose, by contradiction, that x is not basic. Then, by Proposition 4.6, the set of columns in A_{P_x} is not linearly independent. Then, there is

$$\beta \neq 0$$

in $\mathbb{R}^{|P_x|}$ such that $A_{P_x}\beta = 0$. Let ℓ be the index of the component of β with the greatest absolute value, ℓ' the index of the smallest component of x_{P_x}, and let

$$\gamma = \frac{(x_{P_x})_{\ell'}}{|\beta_\ell|}.$$

Then, for every $j = 1, \ldots, |P_x|$, we have

$$(x_{P_x})_j \geq (x_{P_x})_{\ell'} = (x_{P_x})_{\ell'}\frac{|\beta_\ell|}{|\beta_\ell|} \geq (x_{P_x})_{\ell'}\frac{|\beta_j|}{|\beta_\ell|} = \frac{(x_{P_x})_{\ell'}}{|\beta_\ell|}|\beta_j| = \gamma|\beta_j|.$$

Hence,

$$(x_{P_x})_j \geq \gamma\beta_j \quad \text{and} \quad (x_{P_x})_j \geq -\gamma\beta_j.$$

So,

$$(x_{P_x})_j - \gamma\beta_j \geq 0 \quad \text{and} \quad (x_{P_x})_j + \gamma\beta_j \geq 0.$$

Observe that $\gamma\beta \neq 0$. Let $w \in \mathbb{R}^n$ be such that

$$w_{P_x} = \gamma\beta \quad \text{and} \quad w_j = 0 \text{ for every } j \in N \setminus P_x.$$

Then, $w \neq 0$ and

$$x - w \geq 0 \quad \text{and} \quad x + w \geq 0$$

since $x \geq 0$. On the other hand,

$$A(x+w) = Ax + Aw = b + A_{P_x} w_{P_x} = b + A_{P_x} \gamma \beta = b + 0 = b$$

and similarly

$$A(x-w) = b.$$

Therefore, $x + w$ and $x - w$ are admissible vectors. Observe that

$$x = \frac{1}{2}(x+w) + \frac{1}{2}(x-w).$$

Thus, $x = x+w = x-w$ and so $w = 0$ which is a contradiction since we proved before that $w \neq 0$. QED

4.2 Using Basic Vectors

In this section, we prove sufficient conditions to find optimizers. The following result gives a condition that guarantees that, for each admissible vector, there is a basic admissible vector with a better or at least the same objective.

Proposition 4.8

Let P be a standard optimization problem such that the objective map is bounded from below in X_P. Then, for every $z \in X_P$, there exists a basic vector $\tilde{z} \in X_P$ such that $c\tilde{z} \leq cz$.

Proof:

Take $z \in X_P$. Consider the set

$$X_P^z = \{x \in X_P : cx \leq cz\}.$$

Observe that $X_P^z \neq \emptyset$. Among the elements of X_P^z, pick $\tilde{z}$ with the largest possible number of zero components. That is, if $y \in X_P^z$ then

$$|N \setminus P_y| \leq |N \setminus P_{\tilde{z}}|.$$

Consider two cases.

(1) $\tilde{z} = 0$. Then, $P_{\tilde{z}} = \emptyset$. Thus, the set of columns of A with indices in $P_{\tilde{z}}$ is linearly independent since it is the emptyset (see Example 1.26). Hence, by Proposition 4.6, $\tilde{z}$ is a basic vector.

(2) $\tilde{z} \neq 0$. Assume, by contradiction, that the set of columns of A with indices

in $P_{\tilde{z}}$ is linearly dependent. Then, there exists a non-zero vector $v \in \mathbb{R}^{|P_{\tilde{z}}|}$ such that

$$A_{P_{\tilde{z}}} v = 0.$$

Assume that $c_{P_{\tilde{z}}} v < 0$ (the other cases follow in a similar way). Let $w \in \mathbb{R}^n$ be such that

$$\begin{cases} w_j = v_j & \text{whenever } j \in P_{\tilde{z}} \\ 0 & \text{otherwise.} \end{cases}$$

That is, $w_{P_{\tilde{z}}} = v$ and, moreover, $Aw = 0$. Furthermore, $cw < 0$. Consider the family

$$\{y^t\}_{t \in \mathbb{R}_0^+}$$

of vectors in $\mathbb{R}^n$, where

$$y^t = \tilde{z} + tw$$

for each $t \in \mathbb{R}_0^+$. Observe that

$$Ay^t = b,$$

since $A\tilde{z} = b$ and $Aw = 0$. The proof proceeds by cases:

(a) $w \geq 0$. Then, for every $t \geq 0$, the vector $y^t = \tilde{z} + tw$ is admissible, since $y^t \geq 0$, because $\tilde{z} \geq 0$, as $\tilde{z}$ is admissible, and $tw \geq 0$. On the other hand,

$$cy^t = c\tilde{z} + tcw$$

and so the value of $cy^t \to -\infty$ when $t \to +\infty$ because $cw < 0$. Therefore, the objective map is not bounded from below on X_P, contradicting the hypothesis.

(b) There exists $j \in P_{\tilde{z}}$ such that $w_j < 0$. Let

- $u = \max\{-\dfrac{\tilde{z}_\ell}{w_\ell} : w_\ell < 0\}$;

- ℓ^* be such that $u = -\dfrac{\tilde{z}_{\ell^*}}{w_{\ell^*}}$.

(i) $y_k^u = 0$ for $k \in (N \setminus P_{\tilde{z}})$.
It is enough to observe that $y_j^u = 0$ whenever $\tilde{z}_j = 0$ due to the definition of w.
(ii) y^u has one more zero than $\tilde{z}$.
By construction, $y_{\ell^*}^u = 0$. Since $w_{\ell^*} < 0$, then $\ell^* \in P_{\tilde{z}}$. So, $\tilde{z}_{\ell^*} > 0$.
(iii) $y_k^u \geq 0$ for $k \in P_{\tilde{z}} \setminus \{\ell^*\}$.
We have three cases. If $w_k = 0$ then by definition $y_k^u = \tilde{z}_k \geq 0$. Assume that $w_k > 0$. Then, $y_k^u \geq 0$. Assume now that $w_k < 0$. By definition of u we have

$$-\frac{\tilde{z}_k}{w_k} \leq u.$$

Thus, $y_k^u = \tilde{z}_k + uw_k \geq 0$.

Hence, y^u has at least one more zero than $\tilde{z}$ and $y^u \in X_P$. This fact contradicts the choice of $\tilde{z}$ since

$$cy^u = c\tilde{z} + ucw \leq c\tilde{z} \leq cz.$$

Hence, $y^u \in X_P^z$.

Thus, the set of columns of $A_{P_{\tilde{z}}}$ is linearly independent. Hence, by Proposition 4.6, we conclude that $\tilde{z}$ is basic.   QED

The importance of Proposition 4.8 is clear in the next result, called *Basic Minimizer Theorem*.

Theorem 4.1

Consider a standard optimization problem P. Assume that X_P is non-empty and that the objective map is bounded from below in X_P. Then, there is a basic admissible vector in S_P.

Proof:

By Proposition 4.8, for each $x \in X_P$ there exists a basic vector $\tilde{x} \in X_P$ such that $c\tilde{x} \leq cx$. Hence,

$$\inf\{cx : x \in X_P\} = \inf\{c\tilde{x} : x \in X_P\}.$$

Since the set of basic admissible vectors is finite, see Proposition 4.5, the set $\{c\tilde{x} : x \in X_P\}$ is also finite but non-empty. Hence, this set has a minimum. Then, there is a minimizer which is basic.   QED

The constructive nature of this result should be stressed: it provides an algorithm to find a minimizer when the objective map has a lower bound in X_P and $X_P \neq \emptyset$. Indeed, since the set of basic admissible vectors is finite (see Proposition 4.5), it suffices to search for a minimizer in this set. Obviously, this is an inefficient algorithm (see Proposition 7.10).

Example 4.3
Consider the following standard optimization problem:

$$\begin{cases} \min_{x} x_1 \\ x_2 - x_3 = 0 \\ x \geq 0. \end{cases}$$

Observe that any vector $(0, r, r)$ with $r \geq 0$ is a minimizer but only the vector $(0, 0, 0)$ is a basic admissible vector.

Example 4.4

Recall the standard optimimization problem P in Example 4.2. The basic admissible vectors are

$$\left(\frac{5}{3}, \frac{8}{3}, 0, 0\right), \quad (3, 0, 4, 0), \quad (0, 1, 0, 5) \quad \text{and} \quad (0, 0, 1, 6).$$

We now prove that the objective map

$$(x_1, x_2, x_3, x_4) \mapsto -x_1 - 4x_2$$

is bounded from below in X_P. Taking into account the expression of the objective map, it is enough to provide an upper bound for the first two components of each admissible vector. It is immediate that $x_1 \leq 3$ and $x_2 \leq 6$ because $2x_1 + x_2 + x_4 = 6$ and $x_1, x_2, x_4 \geq 0$. Hence, Theorem 4.1 can be used to conclude that there is a basic admissible vector which is a minimizer. Since the minimum value of the objective map among the basic vectors is

$$-\frac{37}{3},$$

which is achieved for the vector

$$\left(\frac{5}{3}, \frac{8}{3}, 0, 0\right),$$

this vector is a minimizer of P.

Although Theorem 4.1 was established for a standard optimization problem, it is clearly useful, by applying the relevant transformations, to canonical optimization problems (Proposition 1.3). The corresponding result is called *Constructive Maximizer Theorem*.

Theorem 4.2

Consider a canonical optimization problem P. Assume that X_P is non-empty and that the objective map is bounded from above in X_P. Then, there is $s \in S_P$ such that $f(s) \in S_{CS(P)}$ is a basic admissible vector.

Proof:

Observe that $CS(P)$ is a standard optimization problem, by Proposition 4.1. By Proposition 1.3, $X_{CS(P)} \neq \emptyset$ since $X_P \neq \emptyset$. Let μ be such that

$$cx \leq \mu$$

for every $x \in X_P$. Let $y \in X_{CS(P)}$. Then,

$$y = \left[\begin{array}{c} x \\ b - Ax \end{array} \right]$$

for some $x \in X_P$. Hence,

$$\left[\begin{array}{cc} -c & 0 \end{array} \right] y = -cx \geq -\mu.$$

Therefore, the objective map of $CS(P)$ is bounded from below in $X_{CS(P)}$. Let $s' \in S_{CS(P)}$ be a basic admissible vector (see Theorem 4.1). Hence, $s'|_n \in S_P$, by Proposition 1.3. QED

4.3 Basic Through Counting

We now provide a characterization of basic vectors that will allow to check whether or not a given admissible vector is basic, by just counting the number of non-zero components of the vector. This characterization holds when the optimization problem is non-degenerate.

Recall by Proposition 4.16 that when $Ax = b$ then b is a linear combination of the columns of A.

Definition 4.4
A standard optimization problem is *non-degenerate* whenever b is not a linear combination of less than m columns of A.

Example 4.5
Recall problem P in Example 1.13. We now show that P is non-degenerate. Because of $m = 2$, we must check whether or not b is a combination of any of the columns of A. Indeed, it is not the case that either

$$\left[\begin{array}{c} 6 \\ 6 \end{array} \right] = \alpha \left[\begin{array}{c} 3 \\ -1 \end{array} \right]$$

for some $\alpha \in \mathbb{R}$, or

$$\left[\begin{array}{c} 6 \\ 6 \end{array} \right] = \alpha \left[\begin{array}{c} -1 \\ 3 \end{array} \right]$$

for some $\alpha \in \mathbb{R}$, or

$$\left[\begin{array}{c} 6 \\ 6 \end{array} \right] = \alpha \left[\begin{array}{c} 1 \\ 0 \end{array} \right]$$

for some $\alpha \in \mathbb{R}$, or

$$\begin{bmatrix} 6 \\ 6 \end{bmatrix} = \alpha \begin{bmatrix} 0 \\ 1 \end{bmatrix}$$

for some $\alpha \in \mathbb{R}$.

The following result is known as the *Non-Degeneracy Lemma*.

Proposition 4.9

Let P be a non-degenerate standard optimization problem and $x \in X_P$. Then,

$$x \text{ is basic } \text{ if and only if } \quad |P_x| = m.$$

Proof:

($\rightarrow$) Assume x is basic. Then, by Definition 4.3 of basic vector, $|P_x| \leq m$. We now show that $|P_x| \geq m$. Indeed, assume, by contradiction, that $|P_x| < m$. Observe that, taking into account Proposition 4.16,

$$Ax = A_{P_x} x_{P_x} + A_{(N \setminus P_x)} x_{(N \setminus P_x)} = A_{P_x} x_{P_x} = b.$$

Hence,

$$\sum_{j \in P_x} a_{\bullet j} x_j = b.$$

Therefore, b is a linear combination of less than m columns of A, contradicting the non-degeneracy hypothesis.

($\leftarrow$) Suppose $|P_x| = m$. Without loss of generality, assume that $P_x = \{1, \dots, m\} = M$. Then, by Proposition 4.16,

$$b = Ax = x_1 a_{\bullet 1} + \cdots + x_m a_{\bullet m}.$$

We show that the set of columns $\{a_{\bullet 1}, \dots, a_{\bullet m}\}$ is linearly independent, which by Proposition 4.6 is equivalent to saying that x is basic. Suppose, by contradiction, that the set of columns $\{a_{\bullet 1}, \dots, a_{\bullet m}\}$ is linearly dependent. Then, there exists a non-zero $v \in \mathbb{R}^m$ such that

$$v_1 a_{\bullet 1} + \cdots + v_m a_{\bullet m} = 0.$$

Observe that

$$(x_1 - \alpha v_1) a_{\bullet 1} + \cdots + (x_m - \alpha v_m) a_{\bullet m} = b,$$

for every $\alpha \in \mathbb{R}$. Take

$$\alpha = \begin{cases} \dfrac{x_i}{v_i} & \text{if } v_i > 0 \\[2mm] -\dfrac{x_i}{v_i} & \text{otherwise,} \end{cases}$$

where $i \in M$ is such that $v_i \neq 0$. Then, $(x_i - \alpha v_i) = 0$. Therefore, it follows that b is a linear combination of less than m columns of A, contradicting the non-degeneracy hypothesis. QED

Hence, in the case of non-degeneracy, every basic admissible vector is non-degenerate (see Definition 4.3). Moreover, for any non-degenerate standard optimization problem, we can find the basic admissible vectors by picking up all combinations of $n - m$ variables on $\{x_1, \ldots, x_n\}$ and solve the system of equations of the problem when those variables are 0.

Example 4.6
Consider the standard optimization problem in Example 1.13. There are

$$\binom{4}{2} = 6$$

possible basic admissible vectors. The basic vectors are found as follows:

- $x_3 = x_1 = 0$: the system

$$\begin{cases} -x_2 = 6 \\ 3x_2 + x_4 = 6 \end{cases}$$

  has the solution $(0, -6, 0, 24)$ which is not a basic admissible vector since $(0, -6, 0, 24) \notin X_P$.

- $x_3 = x_2 = 0$: the system

$$\begin{cases} 3x_1 = 6 \\ -x_1 + x_4 = 6 \end{cases}$$

  has the solution $(2, 0, 0, 8)$ which is a basic admissible vector since $(2, 0, 0, 8) \in X_P$.

- $x_3 = x_4 = 0$: the system

$$\begin{cases} 3x_1 - x_2 = 6 \\ -x_1 + 3x_2 = 6 \end{cases}$$

  has the solution $(3, 3, 0, 0)$ which is a basic admissible vector since $(3, 3, 0, 0) \in X_P$.

- $x_4 = x_1 = 0$: the system

$$\begin{cases} -x_2 + x_3 = 6 \\ 3x_2 = 6 \end{cases}$$

has the solution $(0, 2, 8, 0)$ which is a basic admissible vector since $(0, 2, 8, 0) \in X_P$.

- $x_4 = x_2 = 0$: the system

$$\begin{cases} 3x_1 + x_3 = 6 \\ -x_1 = 6 \end{cases}$$

has the solution $(-6, 0, 24, 0)$ which is not a basic admissible vector since $(-6, 0, 24, 0) \notin X_P$.

- $x_1 = x_2 = 0$: the system

$$\begin{cases} x_3 = 6 \\ x_4 = 6 \end{cases}$$

has the solution $(0, 0, 6, 6)$ which is a basic admissible vector since $(0, 0, 6, 6) \in X_P$.

4.4 Solved Problems and Exercises

Problem 4.1
Consider a standard optimization problem. Assume that the coefficients of the matrix are integer numbers. Show that, if x is a basic admissible vector, then for every $j \in N$,

$$|x_j| \leq m! \, \alpha^{m-1} \beta,$$

where

- $\alpha = \max_{i \in M, j \in N} |a_{ij}|$;
- $\beta = \max_{i \in M} |b_i|$.

Solution:
Let x be a basic admissible vector. Then,

$$Ax = A_B x_B = b$$

where B is a basis for x. Since A_B is a nonsingular matrix,

$$x_B = (A_B)^{-1} b.$$

Observe that, (see Proposition 4.22)

$$(A_B)^{-1} = \frac{1}{\det\ A_B}\, (\operatorname{cof} A_B)^{\mathsf{T}}$$

where

$$\operatorname{cof} A_B = \{(-1)^{i+j} \det\ (A_B)_{i,j}\}_{i,j=1,\dots,m}.$$

Let

$$(A_B)_{i,j} = \{e_{k\ell}\}_{k,\ell=1,\dots,m-1}$$

and P be the set of all permutations of $\{1,\dots,m-1\}$. Then, by the Leibniz's Formula (see Proposition 4.19), we have:

$$\det\ (A_B)_{i,j} = \sum_{p \in P} \operatorname{sgn}(p) \prod_{k=1}^{m-1} e_{kp(k)}.$$

Taking into account that

$$\alpha = \max_{i \in M, j \in N} |a_{ij}|,$$

then

$$\prod_{k=1}^{m-1} |e_{kp(k)}| \le \alpha^{m-1}.$$

Hence,

$$|\det\ (A_B)_{i,j}| \le (m-1)!\ \alpha^{m-1}.$$

Since A_B has integer coefficients and is nonsingular, then $|\det A_B|$ is at least one. Thus,

$$0 < \frac{1}{|\det A_B|} \le 1.$$

Finally,

$$
\begin{aligned}
|(x_B)_j| &= |\sum_{i=1}^{m} ((A_B)^{-1})_{j,i}\, b_i| \\
&\leq \sum_{i=1}^{m} |((A_B)^{-1})_{j,i}|\, |b_i| \\
&= \sum_{i=1}^{m} |\frac{1}{\det A_B} ((\mathrm{cof}\, A_B)^{\mathsf{T}})_{j,i}|\, |b_i| \\
&= \sum_{i=1}^{m} |\frac{1}{\det A_B} (\det (A_B)_{i,j})|\, |b_i| \\
&\leq |\frac{1}{\det A_B}| \sum_{i=1}^{m} (m-1)!\, \alpha^{m-1}\, |b_i| \\
&\leq (m-1)!\, \alpha^{m-1} \sum_{i=1}^{m} |b_i| \\
&\leq (m-1)!\, \alpha^{m-1} \sum_{i=1}^{m} \beta \\
&\leq m!\, \alpha^{m-1}\, \beta,
\end{aligned}
$$

for each $j = 1, \ldots, |B|$.   ◁

Problem 4.2

Let P be a standard optimization problem. Show that:

1. if there is a degenerate basic vector then P is degenerate;

2. if b is a linear combination, with non-negative coefficients, of less than m columns (hence, the problem is degenerate) and the objective map in X_P is bounded from below, then there exists a degenerate basic vector in X_P.

Solution:

(1) Let x be a degenerate basic admissible vector of P. Observe that, by definition of degenerate vector,

$$|P_x| < m.$$

Since,

$$A_{P_x} x_{P_x} = b,$$

then b is a linear combination of less than m columns of matrix A. Therefore, P is degenerate.

(2) Assume that b is a linear combination of less than m columns of A with non-negative coefficients. Let b be such that

$$b = \alpha_{i_1} a_{\bullet i_1} + \cdots + \alpha_{i_k} a_{\bullet i_k}$$

with $k < m$ and $\alpha_{i_\ell} \geq 0$ for $\ell = 1, \ldots, k$. Let $x \in \mathbb{R}^n$ be such that

$$x_j = \begin{cases} \alpha_j & \text{if } j \in \{i_1, \ldots, i_k\} \\ 0 & \text{otherwise.} \end{cases}$$

Hence, $x \in X_P$. Thus, invoking Proposition 4.8, there exists a basic admissible vector $\tilde{x}$ such that $c\tilde{x} \leq cx$. From the proof of that result, it follows that $\tilde{x}$ is such that

$$\tilde{x} \in \{z \in X_P : cz \leq cx\}$$

and no element in the latter set has more zero components. Since $x \in \{z \in X_P : cz \leq cx\}$, the number of zeros of x can not be greater than the number of zeros of $\tilde{x}$. Therefore,

$$|P\tilde{x}| \leq |P_x| = k < m.$$

Hence, $\tilde{x}$ is degenerate.   ◁

Problem 4.3
Let P be the following standard optimization problem:

$$\begin{cases} \min_x \ 2x_1 + x_2 \\ -x_1 - x_2 + x_3 = -1 \\ x_1 + x_2 + x_4 = 3 \\ x_1 - x_2 + x_5 = 1 \\ -x_1 + x_2 + x_6 = 1 \\ x \geq 0. \end{cases}$$

Show that there exists a degenerate basic vector in X_P.

Solution:
Observe that

$$\begin{bmatrix} -1 \\ 3 \\ 1 \\ 1 \end{bmatrix} = \begin{bmatrix} -1 \\ 1 \\ 1 \\ -1 \end{bmatrix} + 2 \begin{bmatrix} 0 \\ 1 \\ 0 \\ 0 \end{bmatrix} + 2 \begin{bmatrix} 0 \\ 0 \\ 0 \\ 1 \end{bmatrix}.$$

Therefore, b is a linear combination of columns with indices 1, 4 and 6, and hence of less than 4 columns. On the other hand, 1 is a lower bound in X_P for the objective map. Thus, by Problem 4.2, we conclude that there is a degenerate basic vector in X_P. Namely

$$\tilde{x} = (0, 1, 0, 2, 0, 2) \in X_P$$

is basic and degenerate.   ◁

Exercise 4.2

Consider the following canonical optimization problem P:

$$\begin{cases} \max_{x} \; 2x_1 + 5x_2 \\ -2x_1 + 2x_2 \leq 2 \\ x_1 \leq 2 \\ x \geq 0. \end{cases}$$

(1) Depict the set of admissible vectors.

(2) Use the map CS to obtain a standard optimization problem.

(3) Find the bases and the basic admissible vectors of $CS(P)$.

(4) The set of maximizers of P is non-empty?

(5) Consider the canonical optimization problem resulting from P by removing $x_1 \leq 2$. What can be said about the set of maximizers?

(6) Consider the canonical optimization problem obtained from P by changing the objective map to $(x_1, x_2) \mapsto -x_1 + x_2$. What can be said about the set of maximizers?

Exercise 4.3

Consider the following standard optimization problems:

$$P_1 = \begin{cases} \min_{x} \; cx \\ Ax = b \\ x \geq 0 \end{cases} \qquad P = \begin{cases} \min_{x} \; cx \\ A'x = b' \\ x \geq 0, \end{cases}$$

where

$$A' = \begin{bmatrix} A \\ c \end{bmatrix} \quad \text{and} \quad b' = \begin{bmatrix} b \\ \inf\{cx : x \in X_{P_1}\} \end{bmatrix}.$$

Assume that $X_{P_1} \neq \emptyset$ and that the objective map is bounded from below on X_{P_1}. Show that $S_{P_1} = S_P$.

Exercise 4.4

Consider the following standard optimization problems

$$P_1 = \begin{cases} \min_{x} cx \\ Ax = b \\ x \geq 0 \end{cases} \qquad P_2 = \begin{cases} \min_{x} cx \\ Ax = b \\ x \geq 0 \\ x \leq \mu, \end{cases}$$

where

- P_1 is restricted;

- the components of c are in $\mathbb{Z}$;

- the objective map is bounded from below on X_{P_1};

- the standard optimization problem

$$P = \begin{cases} \min_{x} cx \\ A'x = b' \\ x \geq 0 \end{cases}$$

  with

$$A' = \begin{bmatrix} A \\ c \end{bmatrix} \quad \text{and} \quad b' = \begin{bmatrix} b \\ \inf\{cx : x \in X_{P_1}\} \end{bmatrix}$$

  is restricted;

- μ is

$$(m+1)! \times \max\{\alpha, \max_{1 \leq j \leq n} |c_j|\}^m \times \max\{\beta, \inf\{cx : x \in X_{P_1}\}\},$$

  where

  - $\alpha = \max_{1 \leq i \leq m, 1 \leq j \leq n} |a_{ij}|$;
  - $\beta = \max_{i \in M} |b_i|$.

(1) Show that $S_{P_1} = S_{P_2}$.

(2) Choose A, b and c so that $X_{P_1} \neq X_{P_2}$.

Exercise 4.5

Show that every basic admissible vector with more than one basis is degenerate.

4.5   Relevant Background

In this section, we provide a brief survey of the relevant linear algebra concepts and results needed for getting the new perspective on linear optimization discussed in this chapter. The interested reader can find more details in [46, 38, 7, 14, 43, 59].

The following result and Proposition 1.12 are crucial for the notion of dimension of a vector space to be well defined.

Proposition 4.10

Any two bases of a vector space over a field K have the same the cardinality.

Definition 4.5

The *dimension* of a vector space V over a field K, denoted by

$$\dim V,$$

is the cardinality of any basis of the vector space.

Example 4.7

The dimension of the vector space

$$\{0\}_K$$

is 0, by Example 1.28.

Example 4.8

The dimension of the vector space $\mathbb{R}^n$ is n taking account the standard basis of $\mathbb{R}^n$ (Definition 1.20).

Proposition 4.11

Let V be a vector space over a field K of dimension n and $\{v^1, \ldots, v^k\} \subseteq V$

be a linearly independent set of vectors in V with $k < n$. Then, there are $v^{k+1}, \ldots, v^n \in V$ such that $\{v^1, \ldots, v^n\}$ is a basis of V.

Proposition 4.12
Let V be a vector space over a field K of dimension n and $V' \subseteq V$ a linearly independent set in V such that $|V'| = n$. Then, V' is a basis for V.

Proposition 4.13
Let V_1 band V_2 be subspaces of a vector space over a field K. Then, $\dim V_1 \leq \dim V_2$.

Proposition 4.14
Let V_1 and V_2 be subspaces of a vector space over a field K. Then,

$$\dim(V_1 + V_2) = \dim(V_1) + \dim(V_2) - \dim(V_1 \cap V_2).$$

Finally, we introduce the concept of orthogonal complement of a subspace, to be used later on.

Definition 4.6
Let V_1 be a subspace of the vector space V over a field K. The *orthogonal complement* of V_1, denoted by

$$V_1^{\perp},$$

is the set $\{v \in V : v \cdot v^1 = 0, \text{ for each } v^1 \in V_1\}$.

Example 4.9
Let

$$V_1 = \{(x_1, x_2) \in \mathbb{R}^2 : x_1 = x_2\}.$$

Then, V_1 is a subspace of $\mathbb{R}^2$ with $\dim V_1 = 1$. Moreover,

$$V_1^{\perp} = \{(x_1, x_2) \in \mathbb{R}^2 : x_1 = -x_2\}.$$

Proposition 4.15
Let V_1 be a subspace of the vector space V over a field K. Then,

- $V_1^{\perp}$ is a subspace of V;

- $\dim V_1^{\perp} = \dim V - \dim V_1$;

- $(V_1^\perp)^\perp = V_1$.

Proposition 4.16
Let A be an $m \times n$-matrix, $x \in \mathbb{R}^n$ and $b \in \mathbb{R}^m$ such that $Ax = b$. Then,

$$b = x_1 a_{\bullet 1} + \cdots + x_n a_{\bullet n}.$$

Definition 4.7
An $m \times m$-matrix A is *nonsingular* whenever

$$\{a_{\bullet 1}, \ldots, a_{\bullet m}\}$$

is linearly independent in $\mathbb{R}^m$.

Example 4.10
Let A be the following matrix

$$\begin{bmatrix} -1 & 1 & 1 \\ 2 & -1 & -\frac{3}{2} \\ 1 & 3 & 1 \end{bmatrix}.$$

We start by observing that the set $\{a_{\bullet 1}, a_{\bullet 2}, a_{\bullet 3}\}$ of columns of A is

$$\left\{ \begin{bmatrix} -1 \\ 2 \\ 1 \end{bmatrix}, \begin{bmatrix} 1 \\ -1 \\ 3 \end{bmatrix}, \begin{bmatrix} 1 \\ -\frac{3}{2} \\ 1 \end{bmatrix} \right\}.$$

We now show that this set is linearly dependent. Assume that

$$\alpha_1 \begin{bmatrix} -1 \\ 2 \\ 1 \end{bmatrix} + \alpha_2 \begin{bmatrix} 1 \\ -1 \\ 3 \end{bmatrix} + \alpha_3 \begin{bmatrix} 1 \\ -\frac{3}{2} \\ 1 \end{bmatrix} = 0.$$

Hence,

$$\begin{cases} -\alpha_1 + \alpha_2 + \alpha_3 & = & 0 \\ 2\alpha_1 - \alpha_2 - \frac{3}{2}\alpha_3 & = & 0 \\ \alpha_1 + 3\alpha_2 + \alpha_3 & = & 0. \end{cases}$$

Thus,

$$\begin{cases} \alpha_1 & = & \frac{1}{2}\alpha_3 \\ 0\alpha_3 & = & 0 \\ \alpha_2 & = & -\frac{1}{2}\alpha_3. \end{cases}$$

Thus, choosing, for instance, $\alpha_3 = 2$, $\alpha_1 = 1$ and $\alpha_2 = -1$ we have

$$\alpha_1 \begin{bmatrix} -1 \\ 2 \\ 1 \end{bmatrix} + \alpha_2 \begin{bmatrix} 1 \\ -1 \\ 3 \end{bmatrix} + \alpha_3 \begin{bmatrix} 1 \\ -\frac{3}{2} \\ 1 \end{bmatrix} = \begin{bmatrix} 0 \\ 0 \\ 0 \end{bmatrix}.$$

Therefore, A is singular.

Proposition 4.17
An $m \times m$-matrix A is nonsingular if and only if there is an $m \times m$-matrix B
such that
$$AB = BA = I,$$
where I is the identity matrix.

We now present the important notion of determinant of a square matrix.
Before that, we need some notation.

Notation 4.4
Let A be an $n \times n$-matrix and $i, j \in N$. We denote by

$$A_{i,j}$$

the $(n-1) \times (n-1)$-matrix obtained from A by removing line i and column j.

We define the determinant of a square matrix following the *Laplace's Expansion Formula*.

Definition 4.8
Let A be an $n \times n$-matrix. The *determinant* of A denoted by

$$\det A$$

is

$$\begin{cases} 1 & \text{if } n = 0 \\ \displaystyle\sum_{j=1}^{n} (-1)^{i+j} a_{ij} \det A_{i,j} & \text{otherwise} \end{cases}$$

for a fixed $i \in M$.

Observe that the definition of $\det A$ is independent of i.

Proposition 4.18
Let A be an $n \times n$-matrix. Then,

$$\det A = \begin{cases} 1 & \text{if } n = 0 \\ \displaystyle\sum_{i=1}^{n} (-1)^{i+j} \, a_{ij} \det \, A_{i,j} & \text{otherwise,} \end{cases}$$

for a fixed $j \in M$.

Definition 4.9
Let A be an $n \times n$-matrix and $i, j \in N$. The (i, j)-*minor* of A, denoted by

$$M_{i,j}^{A},$$

is

$$\det \, A_{i,j}.$$

Moreover, the (i, j)-*cofactor* of A, denoted by

$$C_{i,j}^{A},$$

is $(-1)^{i+j} M_{i,j}^{A}$.

Remark 4.2
Observe that, given an $n \times n$-matrix A, we have

$$\det \, A = \begin{cases} 1 & \text{if } n = 0 \\ \displaystyle\sum_{j=1}^{n} a_{ij} C_{i,j}^{A} & \text{otherwise,} \end{cases}$$

for a fixed $i \in M$.

This remark explains why the Laplace's Expansion Formula is also known as the *Cofactor Expansion*. The following characterization of the determinant is known as the *Leibniz's Formula* (see [7]).

Proposition 4.19
Let A be an $n \times n$-matrix. Then,

$$\det \, A = \sum_{p \in P} \operatorname{sgn}(p) \prod_{i=1}^{n} a_{ip(i)},$$

where P is the set of permutation of N and

$$
\text{sgn}(p) = \begin{cases} 1 & \text{if } p \text{ is an even permutation} \\ -1 & \text{otherwise} \end{cases}
$$

(an even permutation is obtained from the identity permutation as the composition of an even number of exchanges of two elements and similarly for odd permutations).

Proposition 4.20
Let A and B be $n \times n$-matrices and I the identity $n \times n$-matrix. Then,

- det $I = 1$;

- $\det(\alpha A) = \alpha^n \det A$;

- $\det(A \times B) = (\det A) \times (\det B)$;

- det $A^{-1} = \dfrac{1}{\det A}$ whenever det $A \neq 0$.

The following result provides a characterization of a nonsingular matrix via the determinant of A.

Proposition 4.21
An $n \times n$-matrix A is nonsingular if and only if det $A \neq 0$.

Proposition 4.22
Let A be a nonsingular $n \times n$-matrix. Then,

$$
A^{-1} = \frac{1}{\det A} (\text{cof } A)^{\mathsf{T}},
$$

where

$$
\text{cof } A = (C_{i,j}^A)_{i,j=1,\ldots,n}
$$

is the matrix of cofactors of A.

The following result is known as the *Cramer's Rule* (see [14]).

Proposition 4.23
Let A be a nonsingular $n \times n$-matrix and $x, b \in \mathbb{R}^n$ such that $x = A^{-1}b$. Then, for each $j = 1, \ldots, n$,

$$x_j = \frac{\det [A]_b^j}{\det A},$$

where $[A]_b^j$ is the matrix obtained from A by replacing column j by b.

Definition 4.10
The *rank* of a matrix A, denoted by

$$\text{rank}(A),$$

is the number of elements of the largest linearly independent set of rows, which coincides with the number of elements of the largest linearly independent set of columns.

Observe that the rank of an $m \times n$-matrix is at most $\min(m, n)$.

Example 4.11
Let A be the following matrix

$$\begin{bmatrix} 1 & 0 \\ 2 & 3 \\ 1 & 2 \end{bmatrix}.$$

Then, rank(A) is 2 since the set composed by the two columns is linearly independent.

Recall the concept of span of a set of vectors in Definition 1.16.

Definition 4.11
Given an $m \times n$-matrix A, the set

$$\text{span}(A) = \text{span}_{\mathbb{R}^m}(\{a_{\bullet j} : j \in N\})$$

is called the *span* of A.

Proposition 4.24
The span of an $m \times n$-matrix A is a subspace of the vector space $\mathbb{R}^m$ and the dimension of span(A) is rank(A).

It is possible to establish the relationship between the rank and the number of columns of a matrix. Before that, we need to introduce the concept of kernel of a matrix.

Definition 4.12

The *kernel* of an $m \times n$-matrix A, denoted by

$$\ker(A),$$

is the set $\{u \in \mathbb{R}^n : Au = 0\}$.

The following result is known as the *Rank-Nullity Theorem* for matrices.

Theorem 4.3

Given a matrix A, then

$$\text{rank}(A) + \dim \ker(A)$$

is the number of columns of A.

Example 4.12

Observe that

$$\ker \left(\begin{bmatrix} -2 & 1 \\ 2 & -1 \end{bmatrix} \right) = \{(x_1, x_2) \in \mathbb{R}^2 : x_2 = 2x_1\}.$$

Hence, by Theorem 4.3,

$$\text{rank} \left(\begin{bmatrix} -2 & 1 \\ 2 & -1 \end{bmatrix} \right) = 1$$

since

$$\dim \ker \left(\begin{bmatrix} -2 & 1 \\ 2 & -1 \end{bmatrix} \right) = 1$$

and the number of columns is 2.

Proposition 4.25

Let A be an $m \times n$-matrix. Then, $\{a_{\bullet 1}, \ldots, a_{\bullet n}\}^{\perp} = \ker(A)$.

Proposition 4.26

Let V be a k-dimensional subspace of $\mathbb{R}^n$ and B be a basis of $V^\perp$. Then,

$$V = \ker(E),$$

where E is the $(n-k) \times n$ matrix whose lines are the elements of B.

Chapter 5

Geometric View

The main objective of this chapter is to give a geometrical characterization of the basic admissible vectors as vertices of an appropriate convex polyhedron. Capitalizing on this relationship, it is possible to compute optimizers through geometrical techniques. In Section 5.4, the reader can find the relevant concepts and results of affine spaces.

5.1 Admissibility

The goal of this section is to introduce several notions concernbing affine spaces that are relevant to standard optimization problems.

We start by introducing the affine space (see Definition 5.12) useful for optimization.

Proposition 5.1
The pair

$$(\mathbb{R}^n, \Theta),$$

where Θ is the map

$$(u, w) \mapsto w - u,$$

is an affine space over $\mathbb{R}^n$.

Proof:

(1) $\Theta_v : \mathbb{R}^n \to \mathbb{R}^n$ is a bijection for each $v \in \mathbb{R}^n$. Indeed:

(a) Injectivity. Let $u \neq w$. Then, $u - v \neq w - v$. Therefore, $\Theta_v(u) \neq \Theta_v(w)$.

117

(b) Θ_v is an onto map. Let $u \in V$. Then, $\Theta_v(u+v) = u$.

(2) $\Theta(u,z) = \Theta(u,w) + \Theta(w,z)$. Indeed:

$$\Theta(u,z) = z - u = (z-w) + (w-u) = \Theta(w,z) + \Theta(u,w).$$

Hence, $(\mathbb{R}^n, \Theta)$ is an affine space over $\mathbb{R}^n$.   QED

The objective now is to define the smallest affine subspace containing a set.

Definition 5.1
The *smallest affine subspace* of $(\mathbb{R}^n, \Theta)$ *containing* W, denoted by

$$\mathbb{A}(W),$$

is the affine subspace

$$\left(\bigcap_{\{U : U \text{ is an affine subspace of } (\mathbb{R}^n, \Theta) \text{ and } W \subseteq U\}} U \right).$$

This notion is well defined since the intersection of affine subspaces is an affine subspace (Proposition 5.20).

Example 5.1
Observe that $\mathbb{A}(\{u\}) = \{u\}$ since $\{u\}$ is an affine subspace, by Example 5.9. Moreover, $\mathbb{A}(\emptyset) = \emptyset$ since $\emptyset$ is an affine subspace (see Definition 5.13).

We now provide another characterization of $\mathbb{A}(W)$. With this purpose in mind, we define affine combination in $\mathbb{R}^n$.

Definition 5.2
An *affine combination* of points $u, w \in \mathbb{R}^n$ is a point of the form

$$\alpha u + (1 - \alpha)w,$$

for some $\alpha \in \mathbb{R}$.

Observe that every convex combination (see Definition 3.1) is an affine combination but not the other around.

Example 5.2
The point $(3, 6)$ is an affine combination of the points $(2, 3)$ and $(1, 0)$. Indeed:

$$\begin{bmatrix} 3 \\ 6 \end{bmatrix} = 2 \begin{bmatrix} 2 \\ 3 \end{bmatrix} - \begin{bmatrix} 1 \\ 0 \end{bmatrix}.$$

Definition 5.3
A set $W \subseteq \mathbb{R}^n$ is *closed under affine combinations* if

$$\alpha u + (1 - \alpha)w \in U$$

whenever $u, w \in W$ and $\alpha \in \mathbb{R}$.

Exercise 5.1

Show that $W \subseteq \mathbb{R}^n$ is closed under affine combinations if and only if

$$\sum_{i=1}^{k} \alpha_i w^i \in W$$

for every $w^1, \ldots, w^k \in W$ and $\alpha_1, \ldots, \alpha_k \in \mathbb{R}$ such that $\sum_{i=1}^{k} \alpha_i = 1$.

Proposition 5.2
Any subset of $\mathbb{R}^n$ is closed under affine combinations if and only if it is an affine subspace of $(\mathbb{R}^n, \Theta)$.

Proof:

Let $U \subseteq \mathbb{R}^n$.

($\rightarrow$) Assume that U is closed under affine combinations. There are two cases:
(1) $U = \emptyset$. Then, U is an affine subspace of $(\mathbb{R}^n, \Theta)$.
(2) $U \neq \emptyset$. Take $t \in U$. We show that

$$\Theta_t(U) = \{x - t : x \in U\}$$

is a subspace of $\mathbb{R}^n$.
(a) $0 \in \Theta_t(U)$, since $0 = t - t$;
(b) If $v \in \Theta_t(U)$, then $\alpha v \in \Theta_t(U)$.
Assume that $v \in \Theta_t(U)$. By definition of $\Theta_t(U)$, we have $v = x - t$ for some $x \in U$. Then, $\alpha v = \alpha(x - t)$. Thus,

$$\alpha(x - t) = \alpha(x - t) + (1 - \alpha)(t - t) = \alpha x + (1 - \alpha)t - t.$$

Observe that
$$\alpha x + (1 - \alpha)t \in U$$
because U is closed under affine combinations. Hence, $\alpha(x - t) \in \Theta_t(U)$.

(c) If $v^1, v^2 \in \Theta_t(U)$ then $v^1 + v^2 \in \Theta_t(U)$.

Assume that $v^1, v^2 \in \Theta_t(U)$. By definition of $\Theta_t(U)$, there exist $x^1, x^2 \in U$ such that $v^1 = x^1 - t$ and $v^2 = x^2 - t$. Therefore, we must show that

$$v^1 + v^2 = (x^1 + x^2) - (t + t) \in \Theta_t(U).$$

Since

$$\frac{1}{2}(x^1 + x^2) = \left(\frac{1}{2}x^1\right) + \left(\left(1 - \frac{1}{2}\right)x^2\right),$$

then

$$\frac{1}{2}(x^1 + x^2) \in U$$

since U is closed under affine combinations. Hence,

$$\frac{1}{2}(x^1 + x^2) - \frac{1}{2}(t + t) \in \Theta_t(U).$$

Thus,

$$(x^1 + x^2) - (t + t) = 2\left(\frac{1}{2}(x^1 + x^2) - \frac{1}{2}(t + t)\right) \in \Theta_t(U),$$

using (b).

($\leftarrow$) Assume that U is an affine subspace of $(\mathbb{R}^n, \Theta)$. Then, let $t \in U$ be such that

$$\Theta_t(U)$$

is a subspace of $\mathbb{R}^n$. Let $u, w \in U$ and $\alpha \in \mathbb{R}$. Then,

$$\alpha\Theta_t(u) + (1 - \alpha)\Theta_t(w) \in \Theta_t(U);$$

that is,

$$\alpha u + (1 - \alpha)w - t \in \Theta_t(U).$$

Therefore, $\alpha u + (1 - \alpha)w \in U$.   QED

Example 5.3

The set $\{x \in \mathbb{R}^n : x \geq 0\}$ is not closed for affine combinations as the following case shows. Observe that

$$2e^1 - e^2 \notin \{x \in \mathbb{R}^n : x \geq 0\},$$

where e^1 and e^2 are vectors in the canonical basis of $\mathbb{R}^n$ (see Definition 1.20). Therefore, by Proposition 5.2, the set $\{x \in \mathbb{R}^n : x \geq 0\}$ is not an affine subspace of $(\mathbb{R}^n, \Theta)$.

Definition 5.4
The *affine hull* of $W \subseteq \mathbb{R}^n$, denoted by

$$A(W),$$

is the set

$$\left\{ \sum_{i=1}^{k} \alpha_i w^i : w^1, \ldots, w^k \in W, \alpha_1, \ldots, \alpha_k \in \mathbb{R}, \sum_{i=1}^{k} \alpha_i = 1 \right\}.$$

Proposition 5.3
Let $W \subseteq \mathbb{R}^n$. Then
$$A(W) = \mathbb{A}(W).$$

Proof:
($\subseteq$) Let U be an affine subspace containing W. We must show that $A(W) \subseteq U$. Observe that, by Proposition 5.2, U is closed under affine combinations. Since U contains W then U is closed under affine combinations of elements of W.

($\supseteq$) Observe that $W \subseteq A(W)$. Moreover, by Proposition 5.2, $A(W)$ is an affine subspace, since $A(W)$ is closed under affine combinations, by Exercise 5.1. Then, $\mathbb{A}(W) \subseteq A(W)$, by definition of $\mathbb{A}(W)$. QED

The following result states that the set of solutions of a system of equations is an affine subspace (see Definition 5.13) over $(\mathbb{R}^n, \Theta)$.

Proposition 5.4
Let A' be an $m' \times n'$-matrix, $b' \in \mathbb{R}^{m'}$,

$$U = \{x \in \mathbb{R}^{n'} : A'x = b'\}$$

and $t \in U$. Then, $\Theta_t(U) = \ker(A')$. So, U is an affine subspace of $(\mathbb{R}^{n'}, \Theta)$.

Proof:
Indeed:
($\subseteq$) Assume that $v \in \Theta_t(U)$. Then, $v = x - t$, where $x \in U$. Hence, $A'v =$

$A'x - A't = b' - b' = 0$. That is, $v \in \ker(A')$.

($\supseteq$) Let $z \in \ker(A')$. Observe that $z = (z+t) - t$ and $A'(z+t) = A'z + A't = 0 + b' = b'$. Thus, $z + t \in U$. Therefore, $z \in \Theta_t(U)$.   QED

To find a characterization of the affine hull and the dimension of the set of admissible vectors, we need the following auxiliary definition and lemma.

Definition 5.5
Let P be a standard optimization problem where A is an $m \times n$-matrix. The *matrix and the vector of implicit equality restrictions* of P, denoted by

$$A^= \quad \text{and} \quad b^=,$$

are

$$\begin{bmatrix} A \\ I^= \end{bmatrix} \quad \text{and} \quad \begin{bmatrix} b \\ 0 \end{bmatrix},$$

respectively, where $I^=$ is the submatrix of I obtained by removing each line with index in

$$N^> = \{j \in N : x_j > 0 \text{ for some } x \in X_P\}.$$

Example 5.4
Consider the following standard optimization problem P:

$$\begin{cases} \min_x -2x_1 - 5x_2 \\ x_1 + x_2 + x_3 = 1 \\ -x_1 - x_2 + x_4 = -1 \\ x \geq 0. \end{cases}$$

Observe that

$$\left(\frac{1}{2}, \frac{1}{2}, 0, 0\right) \in X_P$$

and so $1, 2 \in \{1, 2, 3, 4\}^>$. Then, $x_3 + x_4 = 0$. Therefore, $x_3 = x_4 = 0$. Hence,

$$\{1, 2, 3, 4\}^> = \{1, 2\}.$$

So,

$$A^= = \begin{bmatrix} 1 & 1 & 1 & 0 \\ -1 & -1 & 0 & 1 \\ 0 & 0 & 1 & 0 \\ 0 & 0 & 0 & 1 \end{bmatrix} \quad \text{and} \quad b^= = \begin{bmatrix} 1 \\ -1 \\ 0 \\ 0 \end{bmatrix}.$$

Exercise 5.2

Let P be a standard optimization problem where A is an $m \times n$-matrix. Show that
$$X_P = \{x \in \mathbb{R}^n : A^= x = b^= \text{ and } x_j \geq 0 \text{ for every } j \in N^>\}.$$

Proposition 5.5

Let P be a standard optimization problem where A is an $m \times n$-matrix with $X_P \neq \emptyset$. Then, there is $x \in X_P$ such that $x_j > 0$ for every $j \in N^>$.

Proof:

Let $w^j \in X_P$ be a vector such that $w_j^j > 0$ and $\alpha_j \in]0, 1[$ for every $j \in N^>$ such that
$$\sum_{j \in N^>} \alpha_j = 1.$$

Let
$$x = \sum_{j \in N^>} \alpha_j w^j.$$

It is immediate that $x_j > 0$ for every $j \in N^>$. Moreover,
$$A^= x = \sum_{j \in N^>} \alpha_j A^= w^j = \left(\sum_{j \in N^>} \alpha_j \right) b^= = b^=.$$

Therefore, by Exercise 5.2, $x \in X_P$. QED

Exercise 5.3

Let $W_1, W_2 \subseteq \mathbb{R}^n$ be such that $W_1 \subseteq W_2$. Show that $A(W_1) \subseteq A(W_2)$.

Proposition 5.6

Let P be a standard optimization problem where A is an $m \times n$-matrix with $X_P \neq \emptyset$. Then
$$A(X_P) = \{y \in \mathbb{R}^n : A^= y = b^=\}.$$

Proof:

($\subseteq$) Observe that
$$A(X_P) \subseteq A(\{y \in \mathbb{R}^n : A^= y = b^=\}),$$

by Exercise 5.3, since $X_P \subseteq \{y \in \mathbb{R}^n : A^=y = b^=\}$ (see Exercise 5.2). The thesis follows because

$$A(\{y \in \mathbb{R}^n : A^=y = b^=\}) = \{y \in \mathbb{R}^n : A^=y = b^=\},$$

by Proposition 5.4 and Proposition 5.3.

($\supseteq$) Take $y \in \{y \in \mathbb{R}^n : A^=y = b^=\}$. Let $x \in X_P$ be such that $x_j > 0$ for every $j \in N^>$ (see Proposition 5.5). If $y = x$ then $y \in X_P \subseteq A(X_P)$. Otherwise, consider two cases. Assume that

$$\{k \in N^> : y_k \neq x_k\} = \emptyset.$$

Then, $y_k = x_k \geq 0$ for every $k \in N^>$. Hence, $y \in X_P \subseteq A(X_P)$, by Exercise 5.2. Otherwise, let

$$z^\beta = \beta y + (1 - \beta)x$$

where

$$\beta = \begin{cases} \displaystyle\max_{k \in \{k \in N^> : x_k \neq y_k\}} \dfrac{x_k}{x_k - y_k} & \text{when } x_k - y_k \leq 0 \text{ for each } k \in N^> \\[2ex] \displaystyle\min_{k \in \{k \in N^> : x_k - y_k > 0\}} \dfrac{x_k}{x_k - y_k} & \text{otherwise.} \end{cases}$$

We start by showing that

$$z^\beta \in X_P,$$

taking into account Exercise 5.2. It is immediate that $A^=z^\beta = b^=$. It remains to prove that $z_j^\beta \geq 0$ for every $j \in N^>$. Let $j \in N^>$. We have two cases:

(1) $x_k - y_k \leq 0$ for each $k \in N^>$. Observe that

$$\{k \in N^> : y_k \neq x_k\} \neq \emptyset.$$

Then,

$$\beta < 0.$$

There are two cases:

(a) $x_j - y_j < 0$. Hence,

$$\begin{aligned} z_j^\beta &= \beta y_j + (1 - \beta)x_j \\ &= \beta(y_j - x_j) + x_j \\ &\geq \tfrac{x_j}{x_j - y_j}(y_j - x_j) + x_j \\ &= 0 \end{aligned}$$

(b) $x_j - y_j = 0$. Then,

$$z_j^\beta = x_j > 0.$$

(2) $x_k - y_k > 0$ for some $k \in N^>$. Then,

$$\beta > 0.$$

There are three cases:

(a) $x_j - y_j < 0$. Thus,

$$\beta(y_j - x_j) > 0.$$

On the other hand,

$$\frac{x_j}{x_j - y_j}(y_j - x_j) < 0.$$

Hence,

$$\beta(y_j - x_j) > \frac{x_j}{x_j - y_j}(y_j - x_j).$$

Therefore,

$$z_j^\beta > 0.$$

(b) $x_j - y_j > 0$. Observe that

$$\beta \leq \frac{x_j}{x_j - y_j}.$$

Thus,

$$\beta(y_j - x_j) \geq \frac{x_j}{x_j - y_j}(y_j - x_j).$$

Therefore,

$$z_j^\beta > 0.$$

(c) $x_j - y_j = 0$. Then,

$$z_j^\beta = x_j > 0.$$

It remains to show that

$$y \in A(X_P).$$

Observe that

$$\beta \neq 0$$

and

$$y = \frac{1}{\beta}z^\beta + \frac{\beta - 1}{\beta}x.$$

Thus, y is an affine combination of x and z^β in X_P. So, $y \in A(X_P)$.   QED

We now define the dimension of a set in the context of affine spaces. Recall the concept of dimension of an affine subspace in Definition 5.15.

Definition 5.6
The *dimension* of $W \subseteq \mathbb{R}^n$, denoted by

$$\dim W,$$

is $\dim \mathbb{A}(W)$.

Remark 5.1
Observe that

$$\dim \emptyset = -1$$

taking into account Definition 5.15 and Example 5.1. Moreover, $\dim \emptyset \leq \dim W$ for every $W \subseteq \mathbb{R}^n$.

Proposition 5.7
Consider the standard optimization problem

$$P = \begin{cases} \min_x cx \\ Ax = b \\ x \geq 0, \end{cases}$$

where A is an $m \times n$-matrix. Then,

$$\dim X_P = \begin{cases} n - \mathrm{rank}(A^=) & \text{if } X_P \neq \emptyset \\ -1 & \text{otherwise.} \end{cases}$$

Proof:
If $X_P = \emptyset$ then $\dim X_P = -1$ (see Remark 5.1). Otherwise,

$$\dim X_P = \dim \mathbb{A}(X_P) = \dim \mathsf{A}(X_P) = \dim \{x \in \mathbb{R}^n : A^= x = b^=\}$$

by Proposition 5.3 and Proposition 5.6. Thus,

$$\dim X_P = \dim \ker(A^=) = n - \mathrm{rank}(A^=),$$

by Proposition 5.4 and Theorem 4.3.   QED

5.2 Optimizers

The main objective of this section is to show that basic admissible vectors are vertices of a relevant convex polyhedron.

We now concentrate on the important notion of hyperplane relevant to define polyhedron and faces.

Definition 5.7

A *hyperplane* in $\mathbb{R}^n$ is an affine subspace of $(\mathbb{R}^n, \Theta)$ with dimension $n - 1$.

Exercise 5.4

Show that any hyperplane is non-empty.

The next result is useful for providing a characterization of hyperplanes.

Proposition 5.8

Let $U \neq \emptyset$ be an affine subspace of $(\mathbb{R}^n, \Theta)$ with dimension k. Then,

$$U = \{x \in \mathbb{R}^n : Ex = Et\},$$

where E is an $(n - k) \times n$-matrix, the lines of E are the vectors of a basis of $\Theta_t(U)^\perp$ and $t \in U$.

Proof:

Observe that, for some $t \in U$, $\Theta_t(U)$ is a subspace of $\mathbb{R}^n$ with dimension k. Thus,

$$\Theta_t(U)^\perp$$

is a subspace of $\mathbb{R}^n$ with dimension $n - k$ (Proposition 4.15). Let

$$B_{\Theta_t(U)^\perp}$$

be a basis of $\Theta_t(U)^\perp$ (Proposition 1.5 and Proposition 1.12). Take

$$E$$

as the matrix $(n - k) \times n$ with the vectors of $B_{\Theta_t(U)^\perp}$ as lines. Therefore, using Proposition 4.15,

$$\{v \in \mathbb{R}^n : Ev = 0\} = (\Theta_t(U)^\perp)^\perp = \Theta_t(U). \quad (\dagger)$$

We are ready to show that

$$U = \{x \in \mathbb{R}^n : Ex = Et\}.$$

($\subseteq$) Let $u \in U$. Then, $u = \Theta_t(u) + t$. Thus, $Eu = E\Theta_t(u) + Et$. Hence, by ($\dagger$), $Eu = Et$.

($\supseteq$) Take $x \in \mathbb{R}^n$ such that $Ex = Et$. Then, $E(x - t) = 0$. Hence, $(x - t) \in \Theta_t(U)$ by ($\dagger$). Therefore, $x \in U$. \hfill QED

Proposition 5.9

Let $H \subseteq \mathbb{R}^n$ be a hyperplane and $t \in H$. Then,

$$H = \{x \in \mathbb{R}^n : v \cdot x = v \cdot t\},$$

where v is the unique vector in a basis of $\Theta_t(H)^{\perp}$.

Proof:

Observe that $\Theta_t(H)$ is a subspace of $\mathbb{R}^n$ with dimension $n - 1$. Then,

$$\Theta_t(H)^{\perp}$$

has dimension 1, by Proposition 4.15. The thesis follows by Proposition 5.8.
QED

We now provide a characterization for the set of solutions of a particular equation to be a hyperplane.

Proposition 5.10

Let $v \in \mathbb{R}^n$ and $r \in \mathbb{R}$. Then, $\{x \in \mathbb{R}^n : v \cdot x = r\}$ is a hyperplane if and only if $v \neq 0$.

Proof:

($\rightarrow$) Assume that $v = 0$. There are two cases:

(1) $r = 0$. Then, $\{x \in \mathbb{R}^n : v \cdot x = r\} = \mathbb{R}^n$ is not a hyperplane, since

$$\dim \{x \in \mathbb{R}^n : v \cdot x = r\} = \dim \mathbb{R}^n = n \neq n - 1.$$

(2) $r \neq 0$. Thus, $\{x \in \mathbb{R}^n : v \cdot x = r\} = \emptyset$ is not a hyperplane, since

$$\dim \{x \in \mathbb{R}^n : v \cdot x = r\} = \dim \emptyset = -1 \neq n - 1.$$

($\leftarrow$) Let H be $\{x \in \mathbb{R}^n : v \cdot x = r\}$, where $v \neq 0$. By Proposition 5.4, H is an affine subspace of $(\mathbb{R}^n, \Theta)$ such that $\Theta_t(H) = \ker([v^{\mathsf{T}}])$ for some $t \in H$. Using, Theorem 4.3,

$$n = \dim \ker([v^{\mathsf{T}}]) + 1$$

since $\text{rank}([v^{\mathsf{T}}]) = 1$. Thus, $\dim H = \dim \ker([v^{\mathsf{T}}]) = n - 1$. Therefore, H is a hyperplane.   QED

Proposition 5.11
Let $H \subseteq \mathbb{R}^n$. Then, H is a hyperplane if and only if

$$H = \{x \in \mathbb{R}^n : v \cdot x = r\}$$

for some non-zero $v \in \mathbb{R}^n$ and $r \in \mathbb{R}$.

Proof:
Immediate by Proposition 5.9 and Proposition 5.10. QED

We now introduce the concept of semi-space.

Definition 5.8
Let $\{x \in \mathbb{R}^n : v \cdot x = r\}$ be a hyperplane. The sets

$$\{x \in \mathbb{R}^n : v \cdot x \geq r\} \quad \text{and} \quad \{x \in \mathbb{R}^n : v \cdot x \leq r\}$$

are the *upper semi-space* and the *lower semi-space* of the hyperplane, respectively.

The following result states an important topological property of semi-spaces.

Proposition 5.12
The semi-spaces of a hyperplane are closed sets.

Proof:
Let $\{x \in \mathbb{R}^n : v \cdot x = r\}$ be a hyperplane. The proof that

$$\{x \in \mathbb{R}^n : v \cdot x \geq r\}$$

is a closed set is similar to the one in Proposition 2.9, by choosing the linear map $x \mapsto v \cdot x$. Similarly for $\{x \in \mathbb{R}^n : v \cdot x \leq r\}$. QED

Example 5.5
The set $\{x \in \mathbb{R}^n : x_j \geq 0\}$ is an upper semi-space of the hyperplane

$$\{x \in \mathbb{R}^n : x_j = 0\}$$

(see Proposition 5.11) and so is closed by Proposition 5.12.

Exercise 5.5

Show that every semi-space is convex.

We now show that intersection preserves convexity. Afterwards, we provide an illustration that it is not always the case that convexity is preserved by union.

Proposition 5.13

The class of convex subsets of $\mathbb{R}^n$ is closed under intersection.

Proof:

Let $\{U^k\}_{k \in K}$ be a family of convex sets and

$$U = \bigcap_{k \in K} U^k.$$

We consider two cases:

(1) $U = \emptyset$. Then, by Exercise 3.1, U is convex.

(2) $U \neq \emptyset$. Let $u, v \in U$ and $\alpha \in [0, 1]$. Then, $u, v \in U^k$ for every $k \in K$, hence, since U^k is convex,

$$(\alpha u + (1 - \alpha)v) \in U^k$$

for every $k \in K$. Therefore, $(\alpha u + (1 - \alpha)v) \in U$.   QED

Exercise 5.6

Show that every hyperplane is convex.

The next example shows that the union of convex sets is not always a convex set.

Example 5.6

Let H_1 and H_2 be the hyperplanes

$$\{x \in \mathbb{R}^2 : v \cdot x = 2\} \quad \text{and} \quad \{x \in \mathbb{R}^2 : v \cdot x = 1\},$$

respectively, where v is $(0, 1)$. Then, H_1 and H_2 are convex sets (see Exercise 5.6). Observe that $(1, 2), (1, 1) \in H_1 \cup H_2$. Then,

$$\left(1, \frac{3}{2}\right)$$

is a convex combination of $(1, 2)$ and $(1, 1)$ but it is not in $H_1 \cup H_2$. Hence, $H_1 \cup H_2$ is not convex.

Definition 5.9
A *convex polyhedron* is a finite intersection of semi-spaces. A *convex polytope* is a bounded convex polyhedron.

Exercise 5.7

Show that a convex polyhedron is a convex set.

We now show that these notions provide a geometric characterization of the set of admissible vectors.

Proposition 5.14
The set of admissible vectors of a standard optimization problem is a convex polyhedron and so is a convex set.

Proof:
Let P be a standard optimization problem with $m \times n$-matrix A. Observe that, for each $j = 1, \ldots, n$,

$$\{x \in \mathbb{R}^n : x_j \geq 0\}$$

is a semi-space of $\mathbb{R}^n$ (see Example 5.5). On the other hand, for every $i = 1, \ldots, m$, the i-th line of matrix A, denoted $a_{i\bullet}$, is non-zero (see Remark 1.2). Thus, for each $i = 1, \ldots, m$,

$$\{x \in \mathbb{R}^n : a_{i\bullet} x = b_i\}$$

is a hyperplane in $\mathbb{R}^n$ (see Proposition 5.10). Hence, for every $i = 1, \ldots, m$,

$$\{x \in \mathbb{R}^n : a_{i\bullet} x \leq b_i\} \text{ and } \{x \in \mathbb{R}^n : a_{i\bullet} x \geq b_i\}$$

are semi-spaces of $\mathbb{R}^n$. Observe that X_P is

$$\bigcap_{j=1}^{n} \{x \in \mathbb{R}^n : x_j \geq 0\} \cap \bigcap_{i=1}^{m} (\{x \in \mathbb{R}^n : a_{i\bullet} x \leq b_i\} \cap \{x \in \mathbb{R}^n : a_{i\bullet} x \geq b_i\})$$

and so is the intersection of $n + 2m$ semi-spaces. Therefore, X_P is a convex polyhedron and, as a consequence, by Exercise 5.7, is a convex set.   QED

Before providing a geometric characterization of the basic admissible vectors, it is necessary to define face.

Definition 5.10
Let X be a convex polyhedron in $\mathbb{R}^n$. A set $F \subseteq X$ is a *face* of X *by the hyperplane* $H = \{x \in \mathbb{R}^n : v \cdot x = r\}$ whenever

$$\begin{cases} v \cdot x = r & \text{for every } x \in F \\ v \cdot x > r & \text{for every } x \in (X \setminus F). \end{cases}$$

A *vertex* of X *by* H is a face of X by H with dimension 0. We say that $x \in X$ is a *vertex* of X whenever there exists an hyperplane H such that x is a vertex of X by H.

The intuition that a vertex is a singleton set is justified by the following result.

Proposition 5.15
Let $U \subseteq \mathbb{R}^n$. Then, dim $U = 0$ if and only if U is a singleton set.

Proof:
($\leftarrow$) Observe that, by Example 4.7 and Example 5.9,

$$\dim \{u\} = \dim \mathbb{A}(\{u\}) = \dim \Theta_u(\{u\}) = \dim \{0\}_{\mathbb{R}} = 0.$$

($\rightarrow$) Assume that dim $U = 0$. Then,

$$\Theta_t(U) = \{0\}_{\mathbb{R}}.$$

Taking into account that Θ_t is a bijection, U is a singleton set.   QED

So, in the sequel, for simplifying the presentation, we may identify each vertex with its unique element.

Example 5.7
Consider the standard optimization problem

$$P = \begin{cases} \min\limits_{x} \; -2x_1 - x_2 \\ 3x_1 - x_2 + x_4 = 6 \\ -x_1 + 3x_2 + x_3 = 6 \\ x \geq 0 \end{cases}$$

(see Example 4.6). Then,

$$v = (3, 3, 0, 0) \in X_P$$

is a vertex of X_P. Indeed, take the hyperplane

$$H = \{x \in \mathbb{R}^4 : [0 \ 0 \ 1 \ 1] \, x = 0\}.$$

Observe that

$$[0 \ 0 \ 1 \ 1] \, v = 0.$$

Hence, $v \in H$. It remains to show that

$$[0 \ 0 \ 1 \ 1] \, y = y_3 + y_4 > 0$$

for every $y \in (X_P \setminus \{v\})$. Let $y \in (X_P \setminus \{v\})$. Observe that either $y_3 \neq 0$ or $y_4 \neq 0$, since otherwise $y = v$. On the other hand, $y_3 \geq 0$ and $y_4 \geq 0$ since y is an admissible vector. Therefore, either $y_3 > 0$ or $y_4 > 0$.

We now provide a geometric characterization of basic admissible vectors.

Proposition 5.16

Let P be a standard optimization problem. Then,

$$x \text{ is a vertex of } X_P \quad \text{iff} \quad x \text{ is a basic vector of } X_P.$$

Proof:

Indeed:

($\rightarrow$) Assume that x is a vertex of X_P. Then, there exist a non-zero $v \in \mathbb{R}^n$ and $r \in \mathbb{R}$ such that $v \cdot x = r$ and $v \cdot y > r$ for every $y \in (X_P \setminus \{x\})$. Consider the standard optimization problem P_1:

$$\begin{cases} \min_{y} v \cdot y \\ Ay = b \\ y \geq 0. \end{cases}$$

Observe that $X_{P_1} = X_P$. Since $v \cdot x = \min\{v \cdot y : y \in X_P\} = r$, then

$$x \in S_{P_1}.$$

Applying Theorem 4.1, because X_P is non-empty and the objective map $y \mapsto v \cdot y$ is bounded from below (by r) in X_P, then there exists a basic admissible vector $\tilde{x} \in X_P$ such that

$$v \cdot \tilde{x} = \min\{v \cdot y : y \in X_P\}.$$

Since $v \cdot y > r$ for every $y \in (X_P \setminus \{x\})$, then

$$\tilde{x} = x.$$

Thus, x is basic.

($\leftarrow$) Assume that x is a basic vector of X_P. Let $B \subseteq N$ be a basis for x. Then, $x_j = 0$ for every $j \in (N \setminus B)$. Let $w \in \mathbb{R}^n$ be such that

$$w_j = \begin{cases} 0 & \text{if } j \in B \\ 1 & \text{otherwise.} \end{cases}$$

Then,

(1) $w \cdot x = 0$ by the definition of w.

(2) $w \cdot y > 0$ for each $y \in X_P \setminus \{x\}$. Observe that

$$w \cdot y = \sum_{j \in N \setminus B} y_j.$$

Then, to show that $w \cdot y > 0$, we must prove that there exists $j \in N \setminus B$ such that

$$y_j > 0.$$

Assume, by contradiction, that $y_j = 0$, for every $j \in N \setminus B$. Hence,

$$y_j = x_j, \text{ for every } j \in N \setminus B.$$

On the other hand, since A_B is nonsingular, the system

$$A_B z_B = b$$

has a unique solution. Therefore,

$$y_B = x_B.$$

So, $y = x$ which contradicts the fact that $y \in X_P \setminus \{x\}$.   QED

To calculate the vertices of X_P, we now provide a characterization of a relevant hyperplane containing a vertex.

Definition 5.11

A hyperplane $\{x \in \mathbb{R}^n : v \cdot x = r\}$ is *m-basic*, for $0 < m < n$, whenever v has m components with value 0 and $n - m$ components with value 1 and r is 0.

Proposition 5.17
Let P be a standard optimization problem with an $m \times n$-matrix. Then, x is a vertex of X_P if and only if x is a vertex of X_P by an m-basic hyperplane.

Proof:
($\rightarrow$) Assume that x is a vertex of X_P. Then, by Proposition 5.16, x is a basic admissible vector. Let B be the corresponding (index) basis. Then, $x_j = 0$ for every $j \in N \setminus B$. Take v such that

$$v_j = \begin{cases} 1 & \text{if } j \in N \setminus B \\ 0 & \text{otherwise.} \end{cases}$$

Hence, $v \cdot x = 0$. Assume by contradiction that there is $y \in X_P \setminus \{x\}$ such that $v \cdot y = 0$. Then, $y_{N \setminus B} = 0$. Moreover,

$$A_B y_B = Ay = b = Ax = A_B x_B.$$

Thus, $y_B = x_B$ since A_B is a nonsingular matrix. So, $y = x$ contradicting the hypothesis.
($\leftarrow$) Immediate by the definition of vertex.   QED

Example 5.8
Consider the standard optimization problem

$$P = \begin{cases} \min_{x} \ -2x_1 - 5x_2 \\ -2x_1 + 2x_2 + x_3 = 2 \\ x_1 + x_4 = 2 \\ x \geq 0. \end{cases}$$

By, Proposition 5.14, X_P is a convex polyhedron. We now find the vertices of X_P.

Taking into account Proposition 5.17, the 2-basic hyperplanes associated to a vertex are of the form
$$\{x \in \mathbb{R}^4 : v \cdot x = 0\},$$

where $v \in \mathbb{R}^4$ is a vector having 2 components with value 0 and the others with value 1. Observe that a possible vertice x of X_P by the hyperplane is such that $x_j = 0$ whenever $v_j = 1$. We have the following candidates for v:

(1) $v = (1, 1, 0, 0)$. We start by investigating if there is $x \in X_P$ such that $v \cdot x = 0$. Then, $x_1 = x_2 = 0$. Moreover, the system of restrictions becomes

$$\begin{cases} x_3 = 2 \\ x_4 = 2 \\ x_3, x_4 \geq 0. \end{cases}$$

Hence, $x = (0, 0, 2, 2) \in X_P$. Furthermore, $v \cdot y > 0$ whenever $y \in X_P \setminus \{x\}$. Hence, x is a vertex of X_P.

(2) $v = (1, 0, 1, 0)$. We start by analyzing if there is $x \in X_P$ such that $v \cdot x = 0$. Then, $x_1 = x_3 = 0$. Moreover, the system of restrictions becomes

$$\begin{cases} x_2 = 1 \\ x_4 = 2 \\ x_2, x_4 \geq 0. \end{cases}$$

Thus, $x = (0, 1, 0, 2) \in X_P$. Furthermore, $v \cdot y > 0$ whenever $y \in X_P \setminus \{x\}$. Hence, x is a vertex of X_P.

(3) $v = (1, 0, 0, 1)$. We start by investigating if there is $x \in X_P$ such that $v \cdot x = 0$. Then, $x_1 = x_4 = 0$. Moreover, the system of restrictions becomes

$$\begin{cases} 2x_2 + x_3 = 2 \\ 0 = 2 \\ x_2, x_3 \geq 0. \end{cases}$$

Hence, there is no vertex of X_P by this hyperplane.

(4) $v = (0, 1, 1, 0)$. We start by analyzing if there is $x \in X_P$ such that $v \cdot x = 0$. Then, $x_2 = x_3 = 0$. Moreover, the system of restrictions becomes

$$\begin{cases} -2x_1 = 2 \\ x_1 + x_4 = 2 \\ x_1, x_4 \geq 0. \end{cases}$$

Hence, there is no vertex of X_P by this hyperplane.

(5) $v = (0, 1, 0, 1)$. We start by investigating if there is $x \in X_P$ such that $v \cdot x = 0$. Then, $x_2 = x_4 = 0$. Moreover, the system of restrictions becomes

$$\begin{cases} -2x_1 + x_3 = 2 \\ x_1 = 2 \\ x_1, x_3 \geq 0. \end{cases}$$

Thus, $x = (2, 0, 6, 0) \in X_P$. Furthermore, $v \cdot y > 0$ whenever $y \in X_P \setminus \{x\}$. Hence, x is a vertex of X_P.

(6) $v = (0, 0, 1, 1)$. We start by analyzing if there is $x \in X_P$ such that $v \cdot x = 0$. Then, $x_3 = x_4 = 0$. Moreover, the system of restrictions becomes

$$\begin{cases} -2x_1 + 2x_2 = 2 \\ x_1 = 2 \\ x_1, x_2 \geq 0. \end{cases}$$

Thus, $x = (2, 3, 0, 0) \in X_P$. Furthermore, $v \cdot y > 0$ whenever $y \in X_P \setminus \{x\}$. Hence, x is a vertex of X_P.

The following result is known as *Vertex Minimizer Theorem*.

Theorem 5.1

Let P be a standard optimization problem. Assume that X_P is non-empty and that the objective map is bounded from below in X_P. Then, there is a vertex of X_P in S_P.

Proof:

By Theorem 4.1, under the assumptions, there is a basic admissible vector in S_P. So, by Proposition 5.16, there is a vertex of X_P in S_P.   QED

5.3 Solved Problems and Exercises

Problem 5.1
Show that
$$\dim \{x \in \mathbb{R}^n : x \geq 0\} = n.$$

Solution:
We start by showing that

$$\mathbb{A}(\{x \in \mathbb{R}^n : x \geq 0\}) = \mathbb{R}^n.$$

By definition, $\mathbb{A}(\{x \in \mathbb{R}^n : x \geq 0\}) \subseteq \mathbb{R}^n$. For the other inclusion let $u \in \mathbb{R}^n$. We consider two cases:

(1) $u \geq 0$. Then, $u \in \{x \in \mathbb{R}^n : x \geq 0\}$. Hence, $u \in \mathbb{A}(\{x \in \mathbb{R}^n : x \geq 0\})$.

(2) $u \not\geq 0$. Then,

$$u = (-1)\overline{u} + 2\underline{u},$$

where $\overline{u}, \underline{u} \in \mathbb{R}^n$ are such that

$$\overline{u}_j = \begin{cases} -u_j & \text{if } u_j < 0 \\ u_j & \text{otherwise} \end{cases}$$

and

$$\underline{u}_j = \begin{cases} 0 & \text{if } u_j < 0 \\ u_j & \text{otherwise} \end{cases}$$

for each $j = 1, \ldots, n$. Since $\overline{u} \geq 0$ and $\underline{u} \geq 0$, we have

$$\overline{u}, \underline{u} \in \{x \in \mathbb{R}^n : x \geq 0\} \subseteq \mathbb{A}(\{x \in \mathbb{R}^n : x \geq 0\}).$$

Hence, u is an affine combination of elements of $\mathbb{A}(\{x \in \mathbb{R}^n : x \geq 0\})$. Thus,

$$u \in \mathbb{A}(\{x \in \mathbb{R}^n : x \geq 0\})$$

by Proposition 5.2.

Therefore, the dimension of $\mathbb{A}(\{x \in \mathbb{R}^n : x \geq 0\})$ is n because $\dim \mathbb{R}^n$ is n, see Example 4.8. Hence, $\dim \{x \in \mathbb{R}^n : x \geq 0\} = n$.   ◁

Problem 5.2
Consider the following standard optimization problem

$$P = \begin{cases} \min_x \ -2x_1 - 5x_2 \\ -2x_1 + 2x_2 + x_3 = 2 \\ x_1 + x_4 = 2 \\ x \geq 0. \end{cases}$$

Show that X_P is a polytope.

Solution:
By Proposition 5.14, X_P is a convex polyhedron. We now show that X_P is bounded. Let $x \in X_P$. Then,

- $x_1 \leq 2$ since $x_1 + x_4 = 2$ and $0 \leq x_4$;

- $x_4 \leq 2$ because $x_1 + x_4 = 2$ and $0 \leq x_1$;

- $x_2 \le 3$ since $-2x_1 + 2x_2 + x_3 = 2$, $0 \le x_1 \le 2$, $0 \le x_3$ and x_2 has the greatest value when x_1 is 2 and x_3 is 0;

- $x_3 \le 6$ because $-2x_1 + 2x_2 + x_3 = 2$, $0 \le x_1 \le 2$, $0 \le x_2$ and x_3 has the greatest value when x_1 is 2 and x_2 is 0.

Therefore,

$$\|x\| = \sqrt{x_1^2 + x_2^2 + x_3^2 + x_4^2}$$
$$\le \sqrt{2^2 + 3^2 + 6^2 + 2^2}.$$

So, X_P is a polytope. ◁

Problem 5.3
Let P be a standard optimization problem and $x \in X_P$. Then, x is a vertex of X_P if and only if for every $y, z \in X_P$, $x = y = z$ whenever there exists $\alpha \in]0,1[$ such that $x = \alpha y + (1 - \alpha)z$.

Solution:
Using Proposition 5.16, x is a vertex of X_P if and only if x is basic vector of X_P. On the other hand, by Proposition 4.7, x is a basic vector of X_P if and only if for every $y, z \in X_P$, $x = y = z$ whenever there exists $\alpha \in]0,1[$ such that $x = \alpha y + (1 - \alpha)z$. ◁

Problem 5.4
Let $j \in N$. Show, using the definition, that

$$\{x \in \mathbb{R}^n : x_j = 0\}$$

is a hyperplane.

Solution:
(1) $\{x \in \mathbb{R}^n : x_j = 0\}$ is an affine subspace of $(\mathbb{R}^n, \Theta)$; that is,

$$V = \Theta_t(\{x \in \mathbb{R}^n : x_j = 0\}) = \{x - t : x \in \mathbb{R}^n \text{ and } x_j = 0\}$$

is a subspace of $\mathbb{R}^n$ with $t \in \{x \in \mathbb{R}^n : x_j = 0\}$. Indeed:
(a) $0 \in V$ since $t - t \in V$.
(b) Assume that $x - t, y - t \in V$. Then, $x, y \in \mathbb{R}^n$ and $x_j = y_j = t_j = 0$. Thus, $x + y - t \in \mathbb{R}^n$, $(x + y - t)_j = 0$. Therefore, $(x + y) - 2t \in V$.

(c) Assume that $x - t \in V$ and $\alpha \in \mathbb{R}$. Then, $x_j = t_j = 0$ and $x \in \mathbb{R}^n$. Then, $(\alpha x)_j = (\alpha t)_j = 0$. Observe that

$$(\alpha x + (1 - \alpha)t)_j = 0.$$

Therefore, $\alpha x + (1 - \alpha)t - t = \alpha(x - t) \in V$.

(2) $\dim\{x \in \mathbb{R}^n : x_j = 0\} = n - 1$. Consider the set

$$E = \{e^1, \ldots, e^{j-1}, e^{j+1}, \ldots, e^n\} \subseteq V,$$

where e^k is as in Definition 1.20 for $k = 1, \ldots, j - 1, j + 1, \ldots, n$. Then, E is a linearly independent set, by Proposition 1.10. Moreover,

$$x - t = \sum_{k \in \{1, \ldots, j-1, j+1, \ldots, n\}} (x_k - t_k)e^k.$$

Hence, E is a basis for V. So, $\dim V = n - 1$.   ◁

Problem 5.5
Let $W \subseteq \mathbb{R}^n$ be a non-empty set, $t \in W$ and V a subspace of $\mathbb{R}^n$ such that $\Theta_t(W) \subseteq V$. Show that there is an affine subspace U such that $\Theta_t(U) = V$ and $W \subseteq U$.

Solution:

Take
$$U = \{v + t : v \in V\}.$$

Then, it is immediate that $\Theta_t(U) = V$. Moreover, $W \subseteq U$ as we show now. Assume that $w \in W$. Then, $\Theta_t(w) = w - t \in V$ since $\Theta_t(W) \subseteq V$. Hence, $w \in U$ because $w = (w - t) + t$.   ◁

Problem 5.6
Let $W \subseteq \mathbb{R}^n$. Show that

$$\Theta_t(\mathbb{A}(W)) = \mathrm{span}_{\mathbb{R}^n}(\Theta_t(W))$$

for every $t \in W$.

Solution:

Let $t \in W$.

($\subseteq$) Assume that $v \in \Theta_t(\mathbb{A}(W))$. Then, $v = u - t$ for some u in $\mathbb{A}(W)$.

Observe that $u \in U'$ for every affine subspace U' such that $W \subseteq U'$. Hence, $v \in \Theta_t(U')$ for every affine subspace U' such that $W \subseteq U'$. The thesis follows immediately since there is, by Problem 5.5, an affine subspace U such that $\Theta_t(U) = \mathrm{span}_{\mathbb{R}^n}(\Theta_t(W))$ and $W \subseteq U$, because $\Theta_t(W) \subseteq \mathrm{span}_{\mathbb{R}^n}(\Theta_t(W))$, by definition.

($\supseteq$) Observe that $\Theta_t(W) \subseteq \Theta_t(\mathbb{A}(W))$ since $W \subseteq \mathbb{A}(W)$. Thus,

$$\mathrm{span}_{\mathbb{R}^n}(\Theta_t(W)) \subseteq \Theta_t(\mathbb{A}(W))$$

since $\mathrm{span}_{\mathbb{R}^n}(\Theta_t(W))$ is the smallest subspace of $\mathbb{R}^n$ containing $\Theta_t(W)$, by definition. ◁

Problem 5.7
Show that the set of minimizers for a standard optimization problem is convex.

Solution:
Let P be a standard optimization problem. If $S_P = \emptyset$ then S_P is a convex set (see Exercise 3.1). Otherwise, assume that $s', s'' \in S_P$ are minimizers. That is, $s', s'' \in X_P$ and

$$cs' = cs'' = \min\{cx : x \in X_P\}.$$

Take $s = \alpha s' + (1 - \alpha)s''$ with $\alpha \in [0, 1]$. Observe that $s \in X_P$ since X_P is a convex set (see Proposition 5.14). On the other hand, for every $x \in X_P$,

$$
\begin{aligned}
cs &= c(\alpha s' + (1 - \alpha)s'') \\
&= \alpha cs' + (1 - \alpha)cs'' \\
&\leq \alpha cx + (1 - \alpha)cx \\
&= cx
\end{aligned}
$$

using the fact that α and $1 - \alpha$ are non-negative. Therefore, $s \in S_P$. ◁

Exercise 5.8

Show that $U \subseteq \mathbb{R}^n$ is convex if and only if U is the set of all linear combinations of the form

$$\sum_{j=1}^{k} \alpha_j u^j,$$

where $k \in \mathbb{N}$ such that $u^j \in U$, $\alpha_j \in \mathbb{R}_0^+$ for $j = 1, \ldots, k$ and

$$\sum_{j=1}^{k} \alpha_j = 1.$$

Exercise 5.9

Let X be a convex polyhedron in $\mathbb{R}^n$. Show that $F \subseteq X$ is a face of X by hyperplane $\{x \in \mathbb{R}^n : v \cdot x = r\}$ if and only if

$$\begin{cases} v \cdot x = r & \text{for every } x \in F \\ v \cdot x < r & \text{for every } x \in (X \setminus F). \end{cases}$$

Exercise 5.10

Consider the canonical optimization problem

$$P = \begin{cases} \max\limits_{x} 2x_1 + 5x_2 \\ -\dfrac{4}{3}x_1 - x_2 \leq -2 \\ 2x_1 + x_2 \leq 10 \\ x \geq 0. \end{cases}$$

Find the vertices of $X_{CS(P)}$.

5.4   Relevant Background

We start by establishing some results about affine spaces. For more details, see [60, 8, 63, 6, 54, 9, 59, 33].

Definition 5.12

Let V be a vector space over a field with characteristic 0.[1] An *affine space* over V is a pair

$$(A, \Theta),$$

where A is a set and $\Theta : A \times A \to V$ is a map such that:

- the map $\Theta_{a_1} : A \to V$, with $\Theta_{a_1}(a) = \Theta(a_1, a)$, is a bijection, for every $a_1 \in A$;

- $\Theta(a_1, a_3) = \Theta(a_1, a_2) + \Theta(a_2, a_3)$, for every $a_1, a_2, a_3 \in A$ (known as *Chasles' Relation*).

The vector space V is the *direction* of A and each $a \in A$ is called a *point*. We now provide some useful identities.

---

[1] See [47] for the definition of characteristic of a field. For instance, $\mathbb{R}$ and $\mathbb{C}$ have characteristic 0.

Proposition 5.18
Let (A, Θ) be an affine space over V. Then,

$$\Theta(a, a) = 0 \quad \text{and} \quad \Theta(a, b) = -\Theta(b, a).$$

Proof:
Both equalities are consequence of Chasles' Relation. Indeed, observe that

$$\Theta(a, a) = \Theta(a, a) + \Theta(a, a).$$

So, $\Theta(a, a)$ is the null vector. Note also that

$$\Theta(a, b) + \Theta(b, a) = \Theta(a, a).$$

Since $\Theta(a, a) = 0$, then $\Theta(a, b) + \Theta(b, a) = 0$.   QED

The following result is known as the *Parallelogram Rule*.

Proposition 5.19
Let (A, Θ) be an affine space over V. Then,

$$\Theta(a, b) = \Theta(a', b') \quad \text{iff} \quad \Theta(a, a') = \Theta(b, b').$$

Proof:
Using Chasles' Relation, we have that

$$\Theta(a, b) = \Theta(a, a') + \Theta(a', b)$$

and

$$\Theta(a', b) = \Theta(a', b') + \Theta(b', b).$$

Thus,

$$\Theta(a, b) = \Theta(a, a') + \Theta(a', b') + \Theta(b', b);$$

that is,

$$\Theta(a, b) - \Theta(a', b') = \Theta(a, a') + \Theta(b', b).$$

Therefore, by Proposition 5.18,

$$\Theta(a, b) - \Theta(a', b') = \Theta(a, a') - \Theta(b, b').$$

So, the equivalence holds.   QED

Affine subspaces play an important role in optimization.

Definition 5.13
Let (A, Θ) be an affine space over V and $U \subseteq A$. We say that U is an *affine subspace* of (A, Θ) if it is either empty or there is $t \in U$ such that

$$\Theta_t(U)$$

is a subspace of V.

Example 5.9
Observe that any singleton subset $\{u\}$ of $\mathbb{R}^n$ is an affine subspace since

$$\Theta_u(\{u\}) = \{0\}_{\mathbb{R}}$$

and $\{0\}_{\mathbb{R}}$ is a subspace by Example 1.24.

The following remark points out that it is possible to generate an affine subspace from a subspace and a vector.

Remark 5.2
Let V be a subspace of $\mathbb{R}^n$ and $t \in \mathbb{R}^n$. Then,

$$\Theta_t^{-1}(V) = \{v + t : v \in V\}$$

is an affine subspace of $(\mathbb{R}^n, \Theta)$.

We now show that the class of affine subspaces of an affine space is closed under intersection.

Proposition 5.20
Let $\{U_j\}_{j \in J}$ be a family of affine subspaces of the affine space (A, Θ) over a vector space V. Then,

$$U = \bigcap_{j \in J} U_j$$

is also an affine subspace of (A, Θ). Moreover. if $U \neq \emptyset$ then

$$\Theta_u(U) = \bigcap_{j \in J} \Theta_u(U_j)$$

for each $u \in U$.

Proof:

Let

$$U = \bigcap_{j \in J} U_j.$$

We have two cases:

(1) $U = \emptyset$. Then, U is an affine subspace of (A, Θ) by definition.

(2) $U \neq \emptyset$. Let $u \in U$. Then, $u \in U_j$ for every $j \in J$. Then,

$$\Theta_u(U_j)$$

is a subspace of V. Hence,

$$\bigcap_{j \in J} \Theta_u(U_j)$$

is a subspace of V (see Proposition 1.6). It remains to prove that

$$\Theta_u(U) = \bigcap_{j \in J} \Theta_u(U_j).$$

($\subseteq$) Since $U \subseteq U_j$ for every $j \in J$, then

$$\Theta_u(U) \subseteq \Theta_u(U_j), \text{ for every } j \in J.$$

So,

$$\Theta_u(U) \subseteq \bigcap_{j \in J} \Theta_u(U_j).$$

($\supseteq$) Let $v \in \bigcap_{j \in J} \Theta_u(U_j)$. Then,

$$v \in \Theta_u(U_j), \text{ for every } j \in J.$$

That is, for each $j \in J$, there is $u_j \in U_j$ such that $\Theta_u(u_j) = v$. Because $\Theta_u : A \to V$ is injective, $u_{j_1} = u_{j_2} = u'$ for every $j_1, j_2 \in J$. So, $u' \in U$. Therefore, $v \in \Theta_u(U)$. QED

On the other hand, the union of affine subspaces is not always an affine subspace.

Example 5.10

Consider the affine subspaces H_1 and H_2 introduced in Example 5.6. We showed that $H_1 \cup H_2$ is not closed under convex combinations. Therefore, $H_1 \cup H_2$ is not closed under affine combinations and so is not an affine subspace by Proposition 5.2.

We now concentrate on defining the dimension of affine spaces and sub-spaces.

Definition 5.14
The *dimension of an affine space* (A, Θ) over V is the dimension of V.

Definition 5.15
Let U be an affine subspace of the affine space (A, Θ) over V. The *dimension* of U, denoted by

$$\dim U$$

is the dimension of the subspace $\Theta_t(U)$ of V when $U \neq \emptyset$ and is set to be -1 otherwise.

Chapter 6

Duality

In this chapter, we introduce the notion of dual of a linear optimization problem (the first duality statement in optimization is attributed to John von Neumann see [21, 62, 61, 27, 23, 26]). The objective is to show that duality provides yet a new technique for finding optimizers of the original problem. We start by providing two main results, the Weak and the Strong Duality Theorems. Then, we concentrate on slacks for the linear optimization problem and prove the Slack Complementarity Theorem. Moreover, a necessary and sufficient condition for the existence (and calculation) of the optimizers is stated, as well as, the Equilibrium Theorem. Finally, we prove a sufficient condition for the existence of a unique minimizer of a standard optimization problem.

6.1 Weak and Strong Duality

Towards defining the dual of a canonical optimization problem, we parameterize the restriction $Ax \le b$ and move it to the objective map as a penalty. This goal is attained by extending the Lagrange multiplier technique used in Analysis (see [31]) for optimizing a map under equality constraints (see [12]).

Definition 6.1
Let $P = (A, b, c)$ be a canonical optimization problem where A is an $m \times n$-matrix and $y \in \mathbb{R}^m$. The *parameterized canonical optimization problem* over P and y is defined as follows:

$$\begin{cases} \max_{x} cx + y^{\mathsf{T}}(b - Ax) \\ x \ge 0. \end{cases}$$

Proposition 6.1

Let $P = (A, b, c)$ be a canonical optimization problem where A is an $m \times n$-matrix and $y \in \mathbb{R}^m$. For every $z \in X_P$, we have

$$\text{if} \quad y \geq 0 \quad \text{then} \quad cz \leq \max_{x \geq 0} cx + y^T (b - Ax).$$

Proof:

Let $z \in X_P$. Hence,

$$cz \leq cz + y^T (b - Az) \leq \max_{x \geq 0} cx + y^T (b - Ax)$$

since $y^T (b - Az) \geq 0$, because $b - Az \geq 0$ and $y \geq 0$.   QED

Thus, each $y \geq 0$ leads to an upper bound for the maximum value of the objective map of P. The main result in duality theory states that the minimum value over y of

$$\max_{x \geq 0} cx + y^T (b - Ax),$$

under the condition $y \geq 0$, is the maximum value of the objective map of P.

Notation 6.1

Let $P = (A, b, c)$ be a canonical optimization problem. We denote by

$$\overline{P}$$

the optimization problem

$$\begin{cases} \min_{y} \max_{x \geq 0} cx + y^\mathsf{T}(b - Ax) \\ y \geq 0. \end{cases}$$

Observe that although the objective map of $\overline{P}$ involves variables x and y, the optimization is done on y (x is a bound variable).

Proposition 6.2

Let $P = (A, b, c)$ be a canonical optimization problem and Q the linear optimization problem

$$\begin{cases} \min_{y} b^\mathsf{T} y \\ A^\mathsf{T} y \geq c^\mathsf{T} \\ y \geq 0. \end{cases}$$

Then, the set of optimizers of Q is equal to the set of optimizers of $\overline{P}$.

Proof:

(1) $S_Q \subseteq S_{\overline{P}}$. Let $r \in S_Q$. Then, $A^\mathsf{T} r \geq c^\mathsf{T}$, $r \geq 0$ and

$$b^\mathsf{T} r \leq b^\mathsf{T} y$$

for every $y \in X_Q$. Let $y \geq 0$. Consider two cases:

(a) $A^\mathsf{T} y \geq c^\mathsf{T}$. Thus,

$$
\begin{aligned}
\max_{x \geq 0} cx + r^\mathsf{T}(b - Ax) &= b^\mathsf{T} r + \max_{x \geq 0}(c - r^\mathsf{T} A)x \\
&= b^\mathsf{T} r \\
&\leq b^\mathsf{T} y \\
&= b^\mathsf{T} y + \max_{x \geq 0}(c - y^\mathsf{T} A)x \\
&= \max_{x \geq 0} cx + y^\mathsf{T}(b - Ax).
\end{aligned}
$$

(b) $A^\mathsf{T} y \not\geq c^\mathsf{T}$. Then, there is $j = 1, \ldots, n$ such that

$$c_j - (A^\mathsf{T} y)_j > 0.$$

Therefore,

$$\max_{x \geq 0}(c - y^\mathsf{T} A)x = +\infty.$$

Then,

$$\max_{x \geq 0} cx + y^\mathsf{T}(b - Ax) = b^\mathsf{T} y + \max_{x \geq 0}(c - y^\mathsf{T} A)x = +\infty.$$

Thus,

$$\max_{x \geq 0} cx + r^\mathsf{T}(b - Ax) \leq \max_{x \geq 0} cx + y^\mathsf{T}(b - Ax).$$

(2) $S_{\overline{P}} \subseteq S_Q$. Let $r \in S_{\overline{P}}$. Then, $r \geq 0$ and

$$\max_{x \geq 0} cx + r^\mathsf{T}(b - Ax) \leq \max_{x \geq 0} cx + y^\mathsf{T}(b - Ax)$$

for every $y \geq 0$. Let $y \in X_Q$. Hence, $y \geq 0$ and $A^\mathsf{T} y \geq c^\mathsf{T}$. We now prove that

$$A^\mathsf{T} r \geq c^\mathsf{T}.$$

Suppose, by contradiction, that $A^\mathsf{T} r \not\geq c^\mathsf{T}$. Then, there is $j = 1, \ldots, n$ such that

$$c_j - (A^\mathsf{T} r)_j > 0.$$

Therefore,

$$\max_{x \geq 0}(c - r^\mathsf{T} A)x = +\infty.$$

Thus,

$$\max_{x \geq 0} cx + r^{\mathsf{T}}(b - Ax) = +\infty.$$

Hence,

$$b^{\mathsf{T}}y + \max_{x \geq 0}(c - y^{\mathsf{T}}A)x = \max_{x \geq 0} cx + y^{\mathsf{T}}(b - Ax) = +\infty.$$

Therefore,

$$c - y^{\mathsf{T}}A \nleq 0$$

which contradicts the hypothesis that $y \in X_Q$.

Thus,

$$
\begin{aligned}
b^{\mathsf{T}}r &= b^{\mathsf{T}}r + \max_{x \geq 0}(c - r^{\mathsf{T}}A)x \\
&= \max_{x \geq 0} cx + r^{\mathsf{T}}(b - Ax) \\
&\leq \max_{x \geq 0} cx + y^{\mathsf{T}}(b - Ax) \\
&= b^{\mathsf{T}}y + \max_{x \geq 0}(c - y^{\mathsf{T}}A)x \\
&= b^{\mathsf{T}}y.
\end{aligned}
$$

So, $r \in S_Q$.   QED

Notation 6.2

In the sequel, we refer to the problem Q introduced in Proposition 6.2, by the *dual* of P. The set of admissible vectors for Q is

$$Y_Q$$

rather that X_Q and the set of optimizers for Q is

$$R_Q$$

rather that S_Q.

Example 6.1

Consider the canonical optimization problem P introduced in Example 1.12. The dual problem Q of P is as follows:

$$
\begin{cases}
\min_{y} 6y_1 + 6y_2 \\
3y_1 - y_2 \geq 2 \\
-y_1 + 3y_2 \geq 1 \\
y \geq 0.
\end{cases}
$$

Exercise 6.1

Let Q be the dual of a canonical optimization problem. Show that if $Y_Q \neq \emptyset$ and the objective map of Q is bounded from below in Y_Q then $R_Q \neq \emptyset$.

The next result, called the *Canonical Weak Duality Theorem*, shows that the value of the objective map of Q for each vector in Y_Q is an upper bound of the value of the objective map of P for any vector in X_P.

Theorem 6.1

Let P be a canonical optimization problem and Q the dual of P. Then,

$$cx \leq b^\mathsf{T} y$$

for every $x \in X_P$ and $y \in Y_Q$.

Proof:

Let $y \in Y_Q$ and $x \in X_P$. Then, $y \geq 0$ and $c^\mathsf{T} \leq A^\mathsf{T} y$. Hence, $c \leq y^\mathsf{T} A$. On the other hand, $x \geq 0$ and $Ax \leq b$. Thus, $cx \leq y^\mathsf{T} Ax$ and $y^\mathsf{T} Ax \leq y^\mathsf{T} b$. Hence, $cx \leq y^\mathsf{T} b$. Thus, $cx \leq b^\mathsf{T} y$, since $b^\mathsf{T} y \in \mathbb{R}$. QED

Proposition 6.3

Let P be a canonical optimization problem and Q the dual of P. Then,

- if $Y_Q \neq \emptyset$ then the objective map of P in X_P has an upper bound;

- if $X_P \neq \emptyset$ then the objective map of Q in Y_Q has a lower bound.

Proof:

We only prove the first statement since the other one is similar. Assume that $Y_Q \neq \emptyset$. There are two cases to consider:

(1) $X_P = \emptyset$, in which case $\{cx : x \in X_P\} = \emptyset$. Hence, each real number is an upper bound of that set.

(2) $X_P \neq \emptyset$, in which case the thesis follows from Theorem 6.1. QED

The next result, called the *Optimality Criterion*, provides a sufficient condition for the existence of optimizers.

Proposition 6.4

Let P be a canonical optimization problem and Q the dual of P. If $x \in X_P$, $y \in Y_Q$ and $cx = b^\mathsf{T}y$ then $x \in S_P$ and $y \in R_Q$.

Proof:

Let $x \in X_P$ and $y \in Y_Q$ be such that $cx = b^\mathsf{T}y$. Then, by Theorem 6.1, the following hold:

1. $cx \leq b^\mathsf{T}y'$ for every $y' \in Y_Q$;

2. $cx' \leq b^\mathsf{T}y$ for every $x' \in X_P$.

In other words:

1. $b^\mathsf{T}y \leq b^\mathsf{T}y'$ for every $y' \in Y_Q$;

2. $cx' \leq cx$ for every $x' \in X_P$.

Then, $y \in R_Q$ and $x \in S_P$.   QED

Example 6.2

Consider the canonical optimization problem P introduced in Example 1.12. Recall the dual problem Q of P in Example 6.1. It is immediate that $(3,3) \in X_P$. We want to use duality to conclude that $(3,3) \in S_P$. To find an admissible vector of Q, observe that the system of linear equations:

$$\begin{cases} 3y_1 - y_2 = 2 \\ -y_1 + 3y_2 = 1 \end{cases}$$

has $\left(\dfrac{7}{8}, \dfrac{5}{8}\right)$ as the solution. Hence, $\left(\dfrac{7}{8}, \dfrac{5}{8}\right) \in Y_Q$ since this vector is also non-negative. Moreover,

$$c \begin{bmatrix} 3 \\ 3 \end{bmatrix} = 9 = b^\mathsf{T} \begin{bmatrix} \frac{7}{8} \\ \frac{5}{8} \end{bmatrix}.$$

Therefore, $(3,3) \in S_P$ and $\left(\dfrac{7}{8}, \dfrac{5}{8}\right) \in R_Q$ by Proposition 6.4.

The following result, called the *Lemma of Existence of Dual*, states that there is always an admissible vector y of the dual problem with objective value as close as envisaged of cs when s is an optimizer of the canonical optimization problem.

Proposition 6.5

Let P be a canonical optimization problem, Q the dual of P and $s \in S_P$. Then, for each $\varepsilon > 0$ there exists $y \in Y_Q$ such that

$$cs \leq b^\mathsf{T} y < (cs + \varepsilon).$$

Proof:

Let $\varepsilon > 0$. Given $s \in S_P$, observe that the system

$$\begin{cases} Ax \leq b \\ cx \geq cs \end{cases}$$

has a non-negative solution (the vector s), whereas the system

$$\begin{cases} Ax \leq b \\ cx \geq cs + \varepsilon \end{cases}$$

has no non-negative solution, since cs is the maximum value of the objective map $x \mapsto cx$ in X_P. Let $\overline{A}$ be the $(m+1) \times n$-matrix

$$\begin{bmatrix} A \\ -c \end{bmatrix}$$

and $\overline{b}^\varepsilon \in \mathbb{R}^{m+1}$ the vector

$$\begin{bmatrix} b \\ -cs - \varepsilon \end{bmatrix}.$$

Denote by P^ε the canonical optimization problem

$$\begin{cases} \max_x c'x \\ \overline{A}x \leq \overline{b}^\varepsilon \\ x \geq 0, \end{cases}$$

where c' is an arbitrary row vector in $\mathbb{R}^n$. Observe that

$$X_{P^\varepsilon} = \emptyset.$$

Then, by Proposition 3.11, we conclude that there are $v \in (\mathbb{R}_0^+)^m$ and $z \in \mathbb{R}_0^+$ such that:

- $(v, z) \neq 0$;

- $(v, z)^\mathsf{T} \overline{A} \geq 0^\mathsf{T}$, and so $A^\mathsf{T} v \geq zc^\mathsf{T}$;

- $(v, z)^\mathsf{T} \overline{b}^\varepsilon < 0$, and so $b^\mathsf{T} v < z(cs + \varepsilon)$.

On the other hand, the problem

$$
\begin{cases}
\max\limits_{x} \; c''x \\[2mm]
\overline{A}x \leq \begin{bmatrix} b \\ -cs \end{bmatrix} \\[2mm]
x \geq 0
\end{cases}
$$

has a non-empty set of admissible vectors. So, by Proposition 3.11, we have

$$
(v, z)^\mathsf{T} \begin{bmatrix} b \\ -cs \end{bmatrix} \geq 0.
$$

Hence,
$$
b^\mathsf{T} v \geq zcs.
$$

Thus,
$$
zcs \leq b^\mathsf{T} v < z(cs + \varepsilon).
$$

Therefore, $z > 0$. Take

$$
y = \frac{1}{z} v.
$$

Then,

- $y \in Y_Q$, since

  - $y \geq 0$, because $v \geq 0$ and $z > 0$;
  - $A^\mathsf{T} y \geq c^\mathsf{T}$, noticing that $A^\mathsf{T} v \geq zc^\mathsf{T}$.

- $cs \leq b^\mathsf{T} y < (cs + \varepsilon)$, since $zcs \leq b^\mathsf{T} v < z(cs + \varepsilon)$.

Hence, for each $\epsilon > 0$, we are able to find a $y \in Y_Q$ in the conditions required by the statement.   QED

As a direct consequence of the above result, if $S_P \neq \emptyset$ then $Y_Q \neq \emptyset$.

Proposition 6.6
Let P be a canonical optimization problem and Q the dual of P. Then,

- if $X_P \neq \emptyset$ and $Y_Q = \emptyset$ then the objective map of P in X_P has no upper bound;

- if $X_P = \emptyset$ and $Y_Q \neq \emptyset$ then the objective map of Q in Y_Q has no lower bound.

Proof:

We only show the first statement since the proof of the second statement is similar. Let $X_P \neq \emptyset$ and $Y_Q = \emptyset$. Assume, by contradiction, that there is an upper bound of the objective map of P in X_P. Then, from Theorem 4.2, we conclude that $S_P \neq \emptyset$. Therefore, invoking Proposition 6.5, $Y_Q \neq \emptyset$, contradicting the hypothesis.   QED

The following result is known as the *Canonical Strong Duality Theorem*.

Theorem 6.2

Let P be a canonical optimization problem and Q the dual of P. Then,

1. if $X_P = \emptyset$ and $Y_Q = \emptyset$ then $S_P = \emptyset$ and $R_Q = \emptyset$;

2. if $X_P \neq \emptyset$ and $Y_Q = \emptyset$ then $S_P = \emptyset$ and $R_Q = \emptyset$ and the objective map of P in X_P has no upper bound;

3. if $X_P = \emptyset$ and $Y_Q \neq \emptyset$ then $S_P = \emptyset$ and $R_Q = \emptyset$ and the objective map of Q in Y_Q has no lower bound;

4. if $X_P \neq \emptyset$ and $Y_Q \neq \emptyset$ then $S_P \neq \emptyset$ and $R_Q \neq \emptyset$ and

$$cs = b^\mathsf{T} r$$

for every $s \in S_P$ and $r \in R_Q$.

Proof:

(1) It is immediate since $S_P \subseteq X_P$ and $R_Q \subseteq Y_Q$.

(2) In this case, Proposition 6.6 implies that the objective map of P in X_P has no upper bound. Therefore, $S_P = \emptyset$. On the other hand, $R_Q = \emptyset$ because Y_Q is empty.

(3) The argument is similar to that of the previous case.

(4) Since $Y_Q \neq \emptyset$, by Theorem 6.1, it follows that the objective map of P in X_P has an upper bound. Hence, by Theorem 4.2, we conclude that $S_P \neq \emptyset$. Similarly, we can show that $R_Q \neq \emptyset$. It remains to prove that $cs = b^\mathsf{T} r$ for every $s \in S_P$ and $r \in R_Q$.

Let s and r be arbitrary elements of S_P and R_Q, respectively. By Theorem 6.1,

$$cs \le b^\mathsf{T} r.$$

We prove that

$$cs \ge b^\mathsf{T} r$$

by contradiction. Indeed, assume that $cs < b^\mathsf{T} r$. Take $\varepsilon = b^\mathsf{T} r - cs > 0$. Then, by Proposition 6.5, there exists $y \in Y_Q$ such that $b^\mathsf{T} y < cs + \varepsilon$. That is, there exists $y \in Y_Q$ such that

$$b^\mathsf{T} y < b^\mathsf{T} r,$$

contradicting the hypothesis that r is a minimizer of Q.   QED

Although the previous results about duality were established for canonical optimization problems, they can be adapted to standard optimization problems.

Exercise 6.2

Show that the *dual of a standard optimization problem* is the linear optimization problem

$$\begin{cases} \max\limits_{y} b^\mathsf{T} y \\ A^\mathsf{T} y \le c^\mathsf{T}. \end{cases}$$

Example 6.3
Consider the standard optimization problem P introduced in Example 1.13. The dual problem Q of P is as follows:

$$\begin{cases} \max\limits_{(y_1,y_2)} 6y_1 + 6y_2 \\ 3y_1 - y_2 \le -2 \\ -y_1 + 3y_2 \le -1 \\ y_1 \le 0 \\ y_2 \le 0. \end{cases}$$

The following exercise establishes the *Standard Weak Duality Theorem*.

Exercise 6.3

Let P be a standard optimization problem and Q the dual of P. Show that

$$b^\mathsf{T} y \le cx$$

for every $x \in X_P$ and $y \in Y_Q$.

Exercise 6.4

Let P be a standard optimization problem and Q the dual of P. Show that for every $x \in X_P$ and $y \in Y_Q$, if

$$b^{\mathsf{T}} y = cx$$

then $x \in S_P$ and $y \in R_Q$.

Example 6.4

Consider the standard optimization problem P introduced in Example 1.13 and recall the dual problem Q of P in Example 6.3. We want to use Exercise 6.4 to conclude that $(3, 3, 0, 0) \in S_P$. Recall that $(3, 3, 0, 0) \in X_P$ (see Example 4.6). To find an admissible vector of Q, observe that the system of linear equations:

$$\begin{cases} 3y_1 - y_2 = -2 \\ -y_1 + 3y_2 = -1 \end{cases}$$

has $\left(-\dfrac{7}{8}, -\dfrac{5}{8}\right)$ as the solution. Hence, $\left(-\dfrac{7}{8}, -\dfrac{5}{8}\right) \in Y_Q$, since this vector is also non-negative. Moreover,

$$c \begin{bmatrix} 3 \\ 3 \\ 0 \\ 0 \end{bmatrix} = -9 = b^{\mathsf{T}} \begin{bmatrix} -\frac{7}{8} \\ -\frac{5}{8} \end{bmatrix}.$$

Hence, $(3, 3, 0, 0) \in S_P$ and $\left(-\dfrac{7}{8}, -\dfrac{5}{8}\right) \in R_Q$, by Exercise 6.4.

The following exercise establishes the *Standard Strong Duality Theorem*.

Exercise 6.5

Let P be a standard optimization problem and Q the dual of P. Show that

1. if $X_P = \emptyset$ and $Y_Q = \emptyset$ then $S_P = \emptyset$ and $R_Q = \emptyset$;

2. if $X_P \neq \emptyset$ and $Y_Q = \emptyset$ then $S_P = \emptyset$ and $R_Q = \emptyset$ and the objective map of P in X_P has no lower bound;

3. if $X_P = \emptyset$ and $Y_Q \neq \emptyset$ then $S_P = \emptyset$ and $R_Q = \emptyset$ and the objective map of Q in Y_Q has no upper bound;

4. if $X_P \neq \emptyset$ and $Y_Q \neq \emptyset$ then $S_P \neq \emptyset$ and $R_Q \neq \emptyset$ and

$$cs = b^{\mathsf{T}} r$$

for every $s \in S_P$ and $r \in R_Q$.

Exercise 6.6

Show that the dual of the linear optimization problem P

$$\begin{cases} \max_{x} 0x \\ Ax \leq b \end{cases}$$

is the problem Q defined as follows:

$$\begin{cases} \min_{y} b^{\mathsf{T}} y \\ A^{\mathsf{T}} y = 0 \\ y \geq 0. \end{cases}$$

Moreover, show that $b^T y \geq 0$ for every $y \in Y_Q$. Finally, show that

1. if $X_P = \emptyset$ and $Y_Q = \emptyset$ then $S_P = \emptyset$ and $R_Q = \emptyset$;

2. if $X_P \neq \emptyset$ and $Y_Q = \emptyset$ then $S_P = \emptyset$ and $R_Q = \emptyset$ and the objective map of P in X_P has no upper bound;

3. if $X_P = \emptyset$ and $Y_Q \neq \emptyset$ then $S_P = \emptyset$ and $R_Q = \emptyset$ and the objective map of Q in Y_Q has no lower bound;

4. if $X_P \neq \emptyset$ and $Y_Q \neq \emptyset$ then $S_P \neq \emptyset$ and $R_Q \neq \emptyset$ and

$$cs = b^{\mathsf{T}} r$$

for every $s \in S_P$ and $r \in R_Q$.

6.2   Complementarity

The objective of this section is to provide techniques for deciding whether or not an admissible vector is an optimizer, using duality and slack variables.

Definition 6.2

Let P be a canonical optimization problem and Q the dual of P. Then,

$$d = x \mapsto b - Ax : \mathbb{R}^n \to \mathbb{R}^m \quad \text{and} \quad e = y \mapsto A^{\mathsf{T}} y - c^{\mathsf{T}} : \mathbb{R}^m \to \mathbb{R}^n$$

are called the *slack maps* for P and Q, respectively.

Note that $d(x) \geq 0$ whenever $x \in X_P$ and $e(y) \geq 0$ whenever $y \in Y_Q$.

Remark 6.1
Observe that when $x \in X_P$ and $d(x)_i = 0$, then the i-th line of A is active in x (see Definition 3.5).

Notation 6.3
When $x \in X_P$, $d(x)$ is the *slack* of x. On the other hand, when $y \in Y_Q$, $e(y)$ is the *slack* of y.

The next result is known as the *Slack Lemma*.

Proposition 6.7
Let P be a canonical optimization problem, Q the dual of P, $x \in X_P$ and $y \in Y_Q$. Then,

- $d(x)^\mathsf{T} y + e(y)^\mathsf{T} x = b^\mathsf{T} y - cx$;
- $d(x)^\mathsf{T} y \geq 0$;
- $e(y)^\mathsf{T} x \geq 0$.

Proof:
(1) $d(x)^\mathsf{T} y + e(y)^\mathsf{T} x = b^\mathsf{T} y - cx$ since

$$
\begin{aligned}
d(x)^\mathsf{T} y + e(y)^\mathsf{T} x &= d(x)^\mathsf{T} y + x^\mathsf{T} e(y) \\
&= (b - Ax)^\mathsf{T} y + x^\mathsf{T} (A^\mathsf{T} y - c^\mathsf{T}) \\
&= b^\mathsf{T} y - x^\mathsf{T} A^\mathsf{T} y + x^\mathsf{T} A^\mathsf{T} y - x^\mathsf{T} c^\mathsf{T} \\
&= b^\mathsf{T} y - x^\mathsf{T} c^\mathsf{T} \\
&= b^\mathsf{T} y - cx.
\end{aligned}
$$

(2) $d(x)^\mathsf{T} y \geq 0$ and $e(y)^\mathsf{T} x \geq 0$, since $x, y, d(x), e(y) \geq 0$.   QED

The following result establishes one of the implications of the Slack Complementarity Theorem.

Proposition 6.8
Let P be a canonical optimization problem, Q the dual of P, $s \in S_P$ and $r \in R_Q$. Then,

$$
d(s)^\mathsf{T} r = 0 \qquad \text{and} \qquad e(r)^\mathsf{T} s = 0.
$$

Proof:

Observe that, by Theorem 6.2, $cs = b^\mathsf{T}r$. So, by Proposition 6.7, the thesis follows.   QED

This result implies that if the i-th component of r is non-zero then the i-th line of A is active in s; that is, $d(s)_i = 0$ (see Remark 6.1). Similarly for $e(r)$.

The following characterization of optimizers is known as the *Slack Complementarity Theorem*.

Theorem 6.3

Let P be a canonical optimization problem, Q the dual of P, $x \in X_P$ and $y \in Y_Q$. Then,

$$x \in S_P \text{ and } y \in R_Q \quad \text{if and only if} \quad d(x)^\mathsf{T}y = 0 \text{ and } e(y)^\mathsf{T}x = 0.$$

Proof:

(1) Assume that $x \in S_P$ and $y \in R_Q$. Then, by Proposition 6.8, $d(x)^\mathsf{T}y = 0$ and $e(y)^\mathsf{T}x = 0$.

(2) Assume that $d(x)^\mathsf{T}y = 0$ and $e(y)^\mathsf{T}x = 0$. Then, by Proposition 6.7, $cx = b^\mathsf{T}y$. Thus, by Proposition 6.4, $x \in S_P$ and $y \in R_Q$.   QED

Theorem 6.3 gives a way to test whether or not $x \in X_P$ is an optimizer. The idea is to solve the system

$$\begin{cases} d(x)^\mathsf{T}y = 0 \\ e(y)^\mathsf{T}x = 0. \end{cases}$$

Then, if there is a solution y such that $y \in Y_Q$, we conclude that $x \in S_P$ and $y \in R_Q$.

Example 6.5

Consider the canonical optimization problem P in Example 1.12 and the dual Q of P in Example 6.1. We check whether or not $(3,3) \in S_P$. Observe that $d(3,3) = 0$. On the other hand, the system

$$e(y)^\mathsf{T}\begin{bmatrix} 3 \\ 3 \end{bmatrix} = \begin{bmatrix} 3y_1 - y_2 - 2 & -y_1 + 3y_2 - 1 \end{bmatrix}\begin{bmatrix} 3 \\ 3 \end{bmatrix} = 0$$

implies that

$$(*) \quad y_1 = -y_2 + \frac{3}{2}.$$

On the other hand, y should be in Y_Q. Hence, substituting y_1 by (*) in

$$3y_1 - y_2 \geq 1,$$

we obtain the following inequality

$$y_2 \leq \frac{7}{8}.$$

Moreover, substituting y_1 by (*) in

$$-y_1 + 3y_2 \geq 2,$$

we obtain the following inequality

$$y_2 \geq \frac{7}{8}.$$

Therefore, we conclude that

$$y_2 = \frac{7}{8}.$$

Thus, using (*) again we get

$$y_1 = \frac{5}{8}.$$

So, by Theorem 6.3, $(3,3) \in S_P$ and $\left(\frac{5}{8}, \frac{7}{8}\right) \in R_Q$.

Exercise 6.7

Consider the following optimization problem:

$$\begin{cases} \max_{x} 2x_1 + x_2 \\ -x_1 + x_2 \leq 1 \\ x_1 + x_2 \leq 3 \\ x_1 - x_2 \leq 1 \\ x \geq 0. \end{cases}$$

Show, using duality, that $(2,1)$ is an optimizer and that $(1,2)$ is not an optimizer.

6.3 Equilibrium

The objective of this section is to provide another way to check whether or not an admissible vector is an optimizer. Moreover, we address the problem of knowing if there is a unique optimizer. Recall Notation 4.3.

Definition 6.3

Let P be a standard optimization problem, Q the dual of P, $x \in \mathbb{R}^n$ and $y \in \mathbb{R}^m$. The system of equations

$$(A_{P_x})^{\mathsf{T}} y = c_{P_x}^{\mathsf{T}}$$

is called the *Equilibrium Condition* for x and y.

The following result, known as the *Equilibrium Theorem*, relates optimizers and the Equilibrium Condition.

Theorem 6.4

Let P be a standard optimization problem, Q the dual of P and $x \in X_P$. Then, the following statements hold:

- if $x \in S_P$ then $R_Q \neq \emptyset$ and the Equilibrium Condition for x and y holds for every $y \in R_Q$.

- if there exists $y \in Y_Q$ such that the Equilibrium Condition holds for x and y then $x \in S_P$.

Proof:

(1) Assume that $x \in S_P$. Then, $S_P \neq \emptyset$. Thus, by Exercise 6.5, $R_Q \neq \emptyset$. Let $y \in R_Q$. Using, Exercise 6.5, $b^{\mathsf{T}} y = cx$. That is,

$$cx - x^{\mathsf{T}} A^{\mathsf{T}} y = 0.$$

Therefore,

$$\sum_{j=1}^{n} x_j (c_j - \sum_{i=1}^{m} y_i a_{ij}) = 0.$$

Hence,

$$(\dagger) \quad \sum_{j \in P_x} x_j (c_j - \sum_{i=1}^{m} y_i a_{ij}) = 0.$$

On the other hand,

$$(c_j - \sum_{i=1}^{m} y_i a_{ij}) \geq 0, \text{ for } j \in P_x$$

since $y \in Y_Q$. Thus, from $(\dagger)$, we conclude that

$$x_j (c_j - \sum_{i=1}^{m} y_i a_{ij}) = 0, \text{ for } j \in P_x.$$

Furthermore,

$$\left(c_j - \sum_{i=1}^{m} y_i a_{ij}\right) = 0, \text{ for } j \in P_x.$$

Hence,

$$(\ddagger) \quad \sum_{j \in P_x} \left(c_j - \sum_{i=1}^{m} y_i a_{ij}\right) = 0.$$

But ($\ddagger$) is the Equilibrium Condition for x and y.

(2) Let $y \in Y_Q$ be such that, for every $j \in N$, if $x_j > 0$, then $\sum_{i=1}^{m} y_i a_{ij} = c_j$. Then, for every $j \in N$

$$x_j\left(c_j - \sum_{i=1}^{m} y_i a_{ij}\right) = 0.$$

Hence, similarly to the proof of (1), $b^\mathsf{T} y = cx$. Therefore, by Exercise 6.4, $x \in S_P$. QED

Exercise 6.8

Let P be a standard optimization problem, Q the dual of P and $x \in X_P$. Show that

- if $R_Q \neq \emptyset$ and the Equilibrium Condition for x and y holds for every $y \in R_Q$ then $x \in S_P$;

- if $x \in S_P$ then there exists $y \in Y_Q$ such that the Equilibrium Condition for x and y holds.

Theorem 6.4 provides a new way to conclude that a given admissible vector x is an optimizer. The idea is to solve the system of equilibrium equations

$$(A_{P_x})^\mathsf{T} y = c_{P_x}^\mathsf{T}$$

and after that check whether or not the solution is an admissible vector of the dual problem.

Example 6.6

Consider the standard optimization problem P in Example 1.13 and its dual Q described in Example 6.3. Take the vector $(3, 3, 0, 0) \in X_P$ (see Example 4.6). Then, $P_{(3,3,0,0)} = \{1, 2\}$ and the Equilibrium Condition is as follows:

$$\begin{bmatrix} 3 & -1 \\ -1 & 3 \end{bmatrix} y = \begin{bmatrix} -2 \\ -1 \end{bmatrix}.$$

The solution of this system of equations is

$$\left(-\frac{7}{8}, -\frac{5}{8}\right).$$

Moreover, this vector is admissible for Q. Indeed, the system of inequalities

$$\begin{bmatrix} 3 & -1 \\ -1 & 3 \\ 1 & 0 \\ 0 & 1 \end{bmatrix} \begin{bmatrix} -\frac{7}{8} \\ -\frac{5}{8} \end{bmatrix} \leq \begin{bmatrix} -2 \\ -1 \\ 0 \\ 0 \end{bmatrix}$$

is satisfied. Hence, by Proposition 6.4, the vector $(3, 3, 0, 0)$ is in S_P.

The following result is known as the *Uniqueness Theorem* and is applied to non-degenerate problems (see Definition 4.4).

Theorem 6.5

Let P be a non-degenerate standard optimization problem, Q the dual of P, $s \in S_P$ and $r \in R_Q$. Assume that:

- s is basic;

- $r^\mathsf{T} a_{\bullet j} < c_j$ for each $j \in (N \setminus P_s)$.

Then, $S_P = \{s\}$ and $R_Q = \{r\}$.

Proof:

(1) $|S_P| = 1$. Assume, by contradiction, that there is $s' \in S_P$ such that

$$s' \neq s.$$

Take $v = s' - s$. Then,

$$Av = A(s' - s) = As' - As = b - b = 0.$$

Hence,

$$\sum_{j \in P_s} (r^\mathsf{T} a_{\bullet j}) v_j + \sum_{j \in (N \setminus P_s)} (r^\mathsf{T} a_{\bullet j}) v_j \;\; = \;\; \sum_{j \in N} (r^\mathsf{T} a_{\bullet j}) v_j$$

$$= \;\; r^\mathsf{T}(Av)$$

$$= \;\; 0.$$

Observe that, for every $j \in P_s$, by Proposition 6.4,

$$r^\mathsf{T} a_{\bullet j} = c_j$$

since $s \in S_P$ and $r \in R_Q$. Therefore,

$$\sum_{j \in P_s} c_j v_j + \sum_{j \in (N \backslash P_s)} (r^\mathsf{T} a_{\bullet j}) v_j = 0.$$

On the other hand, for every $j \in (N \setminus P_s)$, we have:

- $r^\mathsf{T} a_{\bullet j} < c_j$, by hypothesis

- $v_j = s'_j \geq 0$, since $s_j = 0$ and $s' \geq 0$.

Thus,

$$\sum_{j \in P_s} c_j v_j + \sum_{j \in (N \backslash P_s)} c_j v_j \geq 0.$$

Furthermore, as we show in (*), there exists $j \in (N \setminus P_s)$ such that $v_j > 0$. Hence, it follows that

$$\sum_{j \in P_s} c_j v_j + \sum_{j \in (N \backslash P_s)} c_j v_j > 0.$$

So,

$$cv > 0,$$

contradicting $s, s' \in S_P$.

(*) We now show that there exists $j \in (N \setminus P_s)$ such that $v_j > 0$. Assume, by contradiction, that

$$(\dagger) \quad v_j = 0$$

for every $j \in (N \setminus P_s)$. Since the problem is non-degenerate and s is basic, by Proposition 4.9,

$$|P_s| = m.$$

Therefore, by Proposition 4.6,

$$A_{P_s} \text{ is a basis for } s.$$

Thus, A_{P_s} is a nonsingular $m \times m$-matrix. On the other hand, from $Av = 0$, it follows that

$$A_{P_s} v_{P_s} + A_{N \backslash P_s} v_{N \backslash P_s} = 0.$$

Since $v_{N \backslash P_s} = 0$ by ($\dagger$), then

$$A_{P_s} v_{P_s} = 0.$$

Therefore, $v_{P_s} = 0$ because A_{P_s} has a unique solution. Hence, $v = 0$ contradicting $s' \neq s$.

(2) $|R_Q| = 1$. By Proposition 6.4, there is $r \in R_Q$ such that

$$(A_{P_s})^\mathsf{T} r = c_{P_s}^\mathsf{T}.$$

Since matrix A_{P_s} is nonsingular, then $(A_{P_s})^\mathsf{T}$ is also nonsingular. Hence, the Equilibrium Condition

$$(A_{P_s})^\mathsf{T} y = c_{P_s}^\mathsf{T}$$

has a unique solution, which is r. Thus, by Proposition 6.4, $|R_Q| = 1$.   QED

Example 6.7
Consider the standard optimization problem P in Example 1.13 and its dual Q described in Example 6.3. We know, from Example 6.4, that $(3,3,0,0) \in S_P$ and $(-\frac{7}{8}, -\frac{5}{8}) \in R_Q$. We now show that $(3,3,0,0)$ is the unique optimizer of P. Since $(3,3,0,0)$ is basic (see Example 4.6), it remains to check, by Theorem 6.5, that the system of inequalities:

$$\begin{bmatrix} -\dfrac{7}{8} & -\dfrac{5}{8} \end{bmatrix} \begin{bmatrix} 1 & 0 \\ 0 & 1 \end{bmatrix} < \begin{bmatrix} 0 \\ 0 \end{bmatrix}$$

is satisfied. Indeed, this is the case and so, by Theorem 6.5, we conclude that

$$(3,3,0,0)$$

is the unique optimizer of P and that

$$\left(-\frac{7}{8}, -\frac{5}{8} \right)$$

is the unique optimizer of Q.

6.4   Logic of Inequalities

As seen in Section 3.1, Farkas' Lemma (see Proposition 3.6) plays a key role in the theory of linear optimization. Therefore, it is worthwhile to analyze it in more detail, including its pure variant (see Proposition 6.9). The pure variant has a logical interpretation. Considering inequalities as formulas, the pure variant corresponds to saying that a calculus for reasoning with inequalities is sound and complete.

We start by stating the *Pure Variant of Farkas' Lemma*.

Proposition 6.9
Let P be the optimization problem introduced in Exercise 6.6 and Q the dual of P. Then,

$$X_P \neq \emptyset \quad \text{if and only if} \quad \text{for every } w \geq 0, \text{ if } w^\mathsf{T} A = 0^\mathsf{T} \text{ then } w^\mathsf{T} b \geq 0.$$

Proof:
($\rightarrow$) Assume that $X_P \neq \emptyset$. Let $w \in (\mathbb{R}_0^+)^m$ be such that $w^\mathsf{T} A = 0^\mathsf{T}$. Observe that $w \in Y_Q$. Then, by Exercise 6.6,

$$b^\mathsf{T} w \geq 0.$$

($\leftarrow$) Note that $y^\mathsf{T} b = b^\mathsf{T} y \geq 0$ for every $y \in Y_Q$. Thus, $0 \in R_Q$, since $0 \in Y_Q$. Then, by Exercise 6.6, $X_P \neq \emptyset$. QED

To define the *Logic of Inequalities*, we start by introducing the set of formulas.

Definition 6.4
An $m \times n$-*formula* is a system of inequalities of the form $Ax \leq b$, where $b \in \mathbb{R}^m$ and A is an $m \times n$-matrix.

We now introduce a calculus for deriving a formula from another formula.

Definition 6.5
Let $A'x \leq b'$ be an $m' \times n$-formula and $Ax \leq b$ be an $m \times n$-formula. We say that $A'x \leq b'$ is *derived* from $Ax \leq b$, written

$$Ax \leq b \vdash A'x \leq b'$$

if there exists an $m' \times m$-matrix D such that

- D has non-negative components;

- $A' = DA$ and $b' = Db$.

Example 6.8
We show that from the formula

$$\begin{bmatrix} 4 & 3 \\ -3 & 1 \\ 0 & -1 \end{bmatrix} \begin{bmatrix} x_1 \\ x_2 \end{bmatrix} \leq \begin{bmatrix} 10 \\ 6 \\ -4 \end{bmatrix}$$

we can derive the formula

$$[\ 1\quad 0\ ]\begin{bmatrix} x_1 \\ x_2 \end{bmatrix} \leq [\ 0\ ].$$

We must find a matrix D with non-negative coefficients such that

$$D\begin{bmatrix} 4 & 3 \\ -3 & 1 \\ 0 & -1 \end{bmatrix} = [\ 1\quad 0\ ] \quad \text{and} \quad D\begin{bmatrix} 10 \\ 6 \\ -4 \end{bmatrix} = [\ 0\ ].$$

Observe that D has the form:

$$[\ d_1 \quad d_2 \quad d_3\ ]$$

where (d_1, d_2, d_3) is a solution of the system of equations

$$\begin{cases} 4d_1 - 3d_2 + 0d_3 = 1 \\ 3d_1 + d_2 - d_3 = 0 \\ 10d_1 + 6d_2 - 4d_3 = 0. \end{cases}$$

Since $(d_1, d_2, d_3) = (1, 1, 4)$ is a solution, the derivation holds.

Definition 6.6

An $m \times n$-formula $Ax \leq b$ is *inconsistent* when

$$Ax \leq b \vdash 0^\mathsf{T} x \leq -1.$$

Otherwise, is *consistent*.

Exercise 6.9

Show that the following system is inconsistent:

$$\begin{cases} 2x_1 - 5x_2 \geq 0 \\ x_1 \leq -1 \\ x_2 \geq 3. \end{cases}$$

We now present the semantics of the logic. For that, we define satisfaction of a formula in $\mathbb{R}^n$.

Definition 6.7

An $m \times n$-formula $Ax \leq b$ is *satisfied* by $u \in \mathbb{R}^n$, written

$$u \Vdash Ax \leq b,$$

if $Au \leq b$ holds. In this case, we also say that u is a *model* of $Ax \leq b$.

Definition 6.8

The $m \times n$-formula $Ax \leq b$ is *satisfiable* if it has a model.

The following result is called the *Soundness of the Logic of Inequalities Theorem*.

Theorem 6.6

Let $A'x \leq b'$ be an $m' \times n$-formula, $Ax \leq b$ an $m \times n$-formula and $u \in \mathbb{R}^n$. Assume that $u \Vdash (Ax \leq b)$ and $Ax \leq b \vdash A'x \leq b'$. Then, $u \Vdash A'x \leq b'$.

Proof:

The hypothesis $Ax \leq b \vdash A'x \leq b'$ yields the existence of an $m' \times m$-matrix D with non-negative components such that $A' = DA$ and $b' = Db$. We show that $u \Vdash A'x \leq b'$. This amounts to prove that

$$(DAu)_k \leq (Db)_k$$

for $k = 1, \ldots, m'$.

Observe that

- $(DAu)_k = \displaystyle\sum_{i=1}^{m} d_{ki}(Au)_i;$

- $(Db)_k = \displaystyle\sum_{i=1}^{m} d_{ki}b_i.$

From the hypothesis $u \Vdash Ax \leq b$ it follows that $Au \leq b$; that is, $(Au)_i \leq b_i$ for $i = 1, \ldots, m$. Therefore, since each d_{ij} is non-negative, we conclude that

$$(DAu)_k \leq (Db)_k$$

for $k = 1, \ldots, m'$. QED

The following result is called the *Logical Variant of the Farkas' Lemma*.

Proposition 6.10

Let $Ax \le b$ be an $m \times n$-formula. Then, $Ax \le b$ is satisfiable if and only if $Ax \le b$ is consistent.

Proof:

($\rightarrow$) Let $u \in \mathbb{R}^n$ be such that $u \Vdash Ax \le b$. Assume, by contradiction, that $Ax \le b$ is not consistent; that is,

$$(Ax \le b) \vdash (0^\mathsf{T} x \le -1).$$

Then, by Theorem 6.6, $u \Vdash (0^\mathsf{T} x \le -1)$, which is a contradiction.

($\leftarrow$) This implication is proved by contraposition. Assume that $Ax \le b$ is not satisfiable; that is, the system $Ax \le b$ has no solution. Then, by Proposition 6.9, there exists $w \in (\mathbb{R}_0^+)^m$ such that $w^\mathsf{T} A = 0^\mathsf{T}$ and $w^\mathsf{T} b < 0$. Therefore, formula $(0^\mathsf{T} x \le -1)$ is derived from formula $(Ax \le b)$ using matrix

$$\frac{1}{|w^\mathsf{T} b|} w^\mathsf{T}.$$

Hence, by definition, $(Ax \le b)$ is inconsistent.   QED

6.5   Solved Problems and Exercises

Problem 6.1

Let P be the following canonical optimization problem

$$\begin{cases} \max_{x} \dfrac{1}{2}x_1 + x_2 \\ x_1 + 2x_2 \le 12 \\ x_1 \le 4 \\ x \ge 0. \end{cases}$$

(1) Find the dual problem Q of P.

(2) Show that $S_P \ne \emptyset$ and $R_Q \ne \emptyset$.

(3) Define the slack maps.

(4) Show using duality that $(1,1) \notin S_P$.

(5) Show using duality that $(3, \frac{9}{2}) \in S_P$.

Solution:

Observe that

$$A = \begin{bmatrix} 1 & 2 \\ 1 & 0 \end{bmatrix} \quad b = \begin{bmatrix} 12 \\ 4 \end{bmatrix} \quad \text{and} \quad c = \begin{bmatrix} \frac{1}{2} & 1 \end{bmatrix}.$$

(1) The dual Q of P is:

$$\begin{cases} \min\limits_{y} b^T y \\ A^T y \geq c^T \\ y \geq 0 \end{cases} = \begin{cases} \min\limits_{y} 12y_1 + 4y_2 \\ y_1 + y_2 \geq \frac{1}{2} \\ 2y_1 \geq 1 \\ y \geq 0. \end{cases}$$

(2) Observe that $X_P \neq \emptyset$ because, for example, $(1,1) \in X_P$ and $Y_Q \neq \emptyset$ since, for instance, $(\frac{1}{2}, 0) \in Y_Q$. Then, by Theorem 6.2, $S_P \neq \emptyset$ and $R_Q \neq \emptyset$.

(3) The slack maps are defined as follows:

$$d = x \mapsto b - Ax = x \mapsto \begin{bmatrix} 12 - x_1 - 2x_2 \\ 4 - x_1 \end{bmatrix} : \mathbb{R}^2 \to \mathbb{R}^2$$

and

$$e = y \mapsto A^T y - c^T = y \mapsto \begin{bmatrix} y_1 + y_2 - \frac{1}{2} \\ 2y_1 - 1 \end{bmatrix} : \mathbb{R}^2 \to \mathbb{R}^2.$$

(4) We start by showing that there is no $y \in \mathbb{R}^2$ such that

$$\begin{cases} d(1,1)^T y = 0 \\ e(y)^T \begin{bmatrix} 1 \\ 1 \end{bmatrix} = 0 \\ y \in Y_Q. \end{cases}$$

Indeed, the system of equations

$$\begin{cases} 3y_1 + y_2 = 0 \\ 3y_1 + y_2 = \frac{3}{2} \end{cases}$$

does not have a solution. Taking into account (2), let $y \in R_Q$. Therefore, it is not the case that

$$d(1,1)^T y = 0 \quad \text{and} \quad e(y)^T \begin{bmatrix} 1 \\ 1 \end{bmatrix} = 0.$$

Hence, by Theorem 6.3, we conclude that $(1, 1) \notin S_P$.

(5) By Proposition 6.4, for proving that $x = (3, \frac{9}{2}) \in S_P$, it is enough to show that there is $y \in Y_Q$ such that

$$c\left(3, \frac{9}{2}\right) = b^\mathsf{T} y.$$

Take $y = (\frac{1}{2}, 0) \in Y_Q$. Then,

$$b^\mathsf{T} y = 6 = cx.$$

Therefore, x and y are optimizers of P and Q, respectively. ◁

Problem 6.2
Let A be an $m \times n$-matrix and $b \in \mathbb{R}^m$. Show that

$$\{x \in \mathbb{R}^n : Ax \le b\} = \emptyset \quad \text{iff} \quad \{y \in \mathbb{R}^m : y^\mathsf{T} A = 0, y^\mathsf{T} b < 0, y \ge 0\} \ne \emptyset.$$

Solution:
($\rightarrow$) The result is shown by contraposition. Assume that

$$(\dagger) \quad \{y \in \mathbb{R}^m : y^T A = 0, y^\mathsf{T} b < 0, y \ge 0\} = \emptyset.$$

Since $v = 0$ satisfies $v^\mathsf{T} A = 0$ and $v \ge 0$, the linear optimization problem Q

$$\begin{cases} \min_y y^\mathsf{T} b \\ y^\mathsf{T} A = 0 \\ y \ge 0 \end{cases}$$

has v as an optimizer. Indeed, if $y \in Y_Q$, then $y^\mathsf{T} A = 0$ and $y \ge 0$. Therefore, by ($\dagger$), $y^\mathsf{T} b \ge 0$. By Exercise 6.6, we conclude that the problem

$$\begin{cases} \max_x 0x \\ Ax \le b \end{cases}$$

also has an optimizer. So, $\{x \in \mathbb{R}^n : Ax \le b\} \ne \emptyset$.

($\leftarrow$) The result is shown by contradiction. Assume that

- $\{y \in \mathbb{R}^m : y^\mathsf{T} A = 0, y^\mathsf{T} b < 0, y \ge 0\} \ne \emptyset$;
- $\{x \in \mathbb{R}^n : Ax \le b\} \ne \emptyset$.

Then, there exist $y \in \mathbb{R}^m$ and $x \in \mathbb{R}^n$ such that $y^\mathsf{T} A = 0$, $y^\mathsf{T} b < 0$, $y \geq 0$ and $Ax \leq b$. Therefore,

$$0 = y^\mathsf{T} Ax \leq y^\mathsf{T} b < 0,$$

which is a contradiction.   ◁

Problem 6.3

Let P be the following canonical optimization problem

$$\begin{cases} \max\limits_{x} \ -cx \\ -Ax \leq -b \\ x \geq 0, \end{cases}$$

Q the dual of P, $s \in S_P$ and $r \in R_Q$. Show that $s \in S_{P'}$, where P' is the problem

$$\begin{cases} \min\limits_{x} \ cx - r^\mathsf{T} Ax \\ x \geq 0. \end{cases}$$

Solution:

Note that Q is the problem:

$$\begin{cases} \min\limits_{y} \ -b^\mathsf{T} y \\ -A^\mathsf{T} y \geq -c^\mathsf{T} \\ y \geq 0 \end{cases}$$

taking into account Proposition 6.2. By Proposition 6.8,

$$(\dagger) \quad 0 = d(s)^\mathsf{T} r = (-b + As)^\mathsf{T} r$$

and

$$(\ddagger) \quad 0 = e(r)^\mathsf{T} s = (-A^\mathsf{T} r + c^\mathsf{T})^\mathsf{T} s.$$

From ($\ddagger$), we have

$$cs - r^\mathsf{T} As = (c - r^\mathsf{T} A)s = (-A^\mathsf{T} r + c^\mathsf{T})^\mathsf{T} s = 0.$$

It remains to check that $cs - r^\mathsf{T} As \leq cx - r^\mathsf{T} Ax$ for every $x \geq 0$; that is, we have to check that

$$(c - r^\mathsf{T} A)x \geq 0$$

for every $x \geq 0$. Indeed, $c - r^\mathsf{T} A \geq 0$, since $r \in Y_Q$.

Therefore, $s \in S_{P'}$ since $s \geq 0$.   ◁

Exercise 6.10

Consider the optimization problem

$$P = \begin{cases} \max\limits_{x} 2x_1 + x_2 \\ -x_1 + x_2 \leq 1 \\ x_1 + x_2 \leq 3 \\ x_1 - x_2 \leq 1 \\ x \geq 0. \end{cases}$$

Show, using the Equilibrium Condition, that $(2, 1, 2, 0, 0)$ is in $S_{CS(P)}$.

Exercise 6.11

Consider the optimization problem

$$P = \begin{cases} \max\limits_{x} x_1 + 2x_2 \\ x_1 + x_2 \leq 1 \\ 2x_1 + x_2 \leq \frac{3}{2} \\ x \geq 0. \end{cases}$$

Find the dual of P and show directly that both have the same optimizer.

Exercise 6.12

Prove the Canonical Variant of Farkas' Lemma in Proposition 3.11, from the Pure Variant of Farkas' Lemma in Proposition 6.9.

Exercise 6.13

Consider the standard optimization problem

$$P = \begin{cases} \min\limits_{x} 2x_1 + x_2 \\ -x_1 - x_2 + x_3 = -1 \\ x_1 + x_2 + x_4 = 3 \\ x_1 - x_2 + x_5 = 1 \\ -x_1 + x_2 + x_6 = 1 \\ x \geq 0. \end{cases}$$

Let Q be the dual of P. Check whether or not S_P and R_Q are singleton sets.

Exercise 6.14

Let P be a canonical optimization problem. Relate the duals of P and $CS(P)$.

Exercise 6.15

Find the dual of the problem introduced in Example 1.1 and interpret it.

Exercise 6.16

Show that the dual of the dual problem is equivalent to the original problem.

Chapter 7

Complexity

This chapter should be taken as a gentle introduction to complexity theory with examples in linear optimization. In Section 7.1, we present the rigorous definition of the decision problems associated to linear optimization and in Section 7.2, we discuss the representation of vectors and matrices. In Section 7.3, we prove that the standard decision problem is in $\mathcal{NP}$. A deterministic algorithm for getting an optimizer, based on a *brute-force* method, is discussed in Section 7.4. The complexity of this algorithm is shown not to be polynomial. Finally, in the relevant background section, we introduce the main notions and techniques that are necessary for assessing the computational efficiency of an algorithm.

7.1 Optimization Decision Problem

In this section, we present the concepts that are needed to formulate the optimization decision problem. As introduced in Section 7.6, a decision problem is a map $h : \mathcal{D} \to \{0, 1\}$, where $\mathcal{D}$ is a denumerable set. So, we restrict our problems to have rational components.

Notation 7.1
For each $m, n \in \mathbb{N}^+$ with $m < n$, we denote by

$$D_{mn}^{SF},$$

the set of triples (A, b, c) such that:

- A is a non-empty $m \times n$ matrix with rational coefficients;

- $c \in \mathbb{Q}^n$;

- $b \in \mathbb{Q}^m$ such that $b_i \neq 0$ for each $i = 1, \ldots, m$.

Observe that each $(A, b, c) \in D_{mn}^{SF}$ can be seen, according to Notation 1.10, as a standard optimization problem.

Notation 7.2

Furthermore, let

$$D^{SF} = \bigcup_{\substack{m, n \in \mathbb{N}^+ \\ m < n}} D_{mn}^{SF};$$

$$\mathbb{Q}^{\cup} = \bigcup_{n \in \mathbb{N}^+} \mathbb{Q}^n.$$

The following result tells us that there exists an optimizer with rational components for a standard optimization problem provided that the problem has rational numbers as parameters and a non-empty set of optimizers.

Proposition 7.1

Let $P \in D_{mn}^{SF}$ be a standard optimization problem such that $S_P \neq \emptyset$. Then, $S_P \cap \mathbb{Q}^n \neq \emptyset$.

Proof:

Since $S_P \neq \emptyset$, then $x \mapsto cx$ is bounded from below in X_P and $X_P \neq \emptyset$. Hence, by Theorem 4.1, there is a basic admissible vector $s \in S_P$. Then, there is $B \subset N$ such that A_B is nonsingular, $s_{N \setminus B} = 0$, and s_B is the unique solution of

$$A_B = b.$$

Thus, by Cramer's Rule (see Proposition 4.23), each component of s is given by

$$s_j = \frac{\det (A_B)_b^j}{\det A_B}.$$

Clearly, each s_j is in $\mathbb{Q}$ since $\det (A_B)_b^j$ and $\det A_B$ are rational numbers as can be seen by the Leibniz's Formula (see Proposition 4.19).   QED

We now introduce the relevant optimization decision problem.

Definition 7.1

The *Standard Optimization Decision Problem* is the map

$$\text{SDP} : \text{D}^{\text{SF}} \to \{0, 1\}$$

such that

$$\text{SDP}(P) = 1 \quad \text{if and only if} \quad S_P \neq \emptyset$$

for every standard optimization problem P. Similarly for the *Canonical Optimization Decision Problem*

$$\text{CDP} : \text{D}^{\text{SF}} \to \{0, 1\}.$$

7.2 Representation

Before analyzing the complexity of the standard optimization decision problem, we need to discuss the representation of vectors and matrices. Recall the representation of natural and rational numbers in Section 7.6.

Notation 7.3

An n-dimensional vector w of rational numbers is represented by the following string of bits

$$\hat{w} = \hat{w}_1 \, 010 \, \cdots \, 010 \, \hat{w}_n,$$

where $\hat{w}_j$ is the encoding of the rational number w_j, as explained in Remark 7.3, for $j = 1, \ldots, n$. Moreover, we choose 010 as the separator of the encodings of the rational numbers. An $m \times n$-matrix A of rational numbers is represented by

$$\hat{A} = \hat{a}_{1\bullet} \, 011 \, \cdots \, 011 \, \hat{a}_{m\bullet},$$

where $\hat{a}_{i\bullet}$ is the representation of the i-th row of the matrix. Moreover, $(A, b, c) \in \text{D}^{\text{SF}}$ is represented by

$$\widehat{(A, b, c)} = \hat{A} \, 100 \, \hat{b} \, 100 \, \hat{c}.$$

The reader may think of alternative ways of representing vectors and matrices of arbitrary dimensions and ponder on the fact that any reasonably economic representation would serve the purpose at hand.[1]

The representation adopted here is not minimal since it contains some acceptable redundancy that simplifies the computation of the dimension.

---

[1]Redundancy is allowed as long as it does not grow faster then polynomial on the sum of the sizes of the entries of the vector and of the matrix.

input : (k, u)

1. scan k and $\hat{u}$ simultaneously until one ends;

2. if the scan of $\hat{u}$ ends before the scan of k then return 1;

3. if the scan of k ends before the scan of $\hat{u}$ then return 0;

4. return 1.

Figure 7.1: An algorithm for comparing sizes.

In the sequel, it is useful to know the complexity of comparing the size of a vector with a given natural number presented in unary notation (see Definition 7.11).

Proposition 7.2
There is an algorithm for the decision problem

$$\text{sizeGE} : \{1\}^* \times \mathbb{Q}^\cup \rightarrow \{0, 1\}$$

such that $\text{sizeGE}(k, u) = 1$ if and only if the size of k is greater than or equal to the size in bits of u, which is polynomial on the first argument.

Proof:
It is immediate that the algorithm in Figure 7.1 computes sizeGE and is polynomial on the first argument.   QED

Notation 7.4
In the sequel, we denote the algorithm presented in Figure 7.1 by $\mathfrak{a}_{\text{sizeGE}}$.

For the next result, see Definition 7.5.

Proposition 7.3
There exists a polynomial algorithm for $\dim : \mathbb{Q}^\cup \rightarrow \mathbb{N}^+$.

Observe that, for the chosen representation of vectors and matrices, for any non-empty matrix A and vector w,

$$\lvert w \rvert = \left(\sum_{i=1}^{\dim w} (3 + \lvert w_i \rvert) \right) - 3$$

and

$$\lvert A \rvert = \left(\sum_{i=1}^{m} (3 + \lvert a_{i\bullet} \rvert) \right) - 3,$$

where m is the number of rows of matrix A. Furthermore, for each $(A, b, c) \in D^{\mathrm{SF}}$,

$$\lvert (A, b, c) \rvert = \lvert A \rvert + \lvert b \rvert + \lvert c \rvert + 6.$$

In the sequel, it also becomes handy to use

$$\lvert (A, b) \rvert = \lvert A \rvert + \lvert b \rvert + 3$$

and

$$\lvert (A, c) \rvert = \lvert A \rvert + \lvert c \rvert + 3.$$

The following result establishes a bound on the size of the determinant of a square matrix with rational entries.

Proposition 7.4

For every non-empty square matrix A with rational entries,

$$\lvert \det A \rvert \le 3 \lvert A \rvert^4.$$

Proof:

First, we show that, for every square matrix Z with integer entries represented as rational numbers with unit denominator

$$\lvert \det Z \rvert \le 2 \lvert Z \rvert^2.$$

To this end, we prove by induction on the number m of rows of Z that

$$\lvert \det Z \rvert \le 2m \lvert Z \rvert.$$

(Basis) In this case, $\det Z = z_{11}$. Thus, $\lvert \det Z \rvert = \lvert Z \rvert < 2 \lvert Z \rvert$.

(Step) Recall Laplace's Expansion Formula for $j = 1$ (see Definition 4.8),

$$\det Z = \sum_{i=1}^{m} (-1)^{i+1} z_{i1} \det Z_{i,1}.$$

By the induction hypothesis,

$$|\det\ Z_{i,1}| \leq 2(m-1)|Z_{i,1}|$$

and, so, since $|z_{i1}|, |Z_{i,1}| \leq |Z|$ and the size of a product is bounded by the sum of the size of the factors,

$$|(-1)^{i+1} z_{i1} \det\ Z_{i,1}| = |z_{i1} \det\ Z_{i,1}| \leq |Z| + 2(m-1)|Z_{i,1}| \leq 2m|Z| - |Z|.$$

Therefore, since the size of a sum of m integer terms is bounded by the sum of m and a bound of the size of each term,

$$|\det\ Z| \leq 2m|Z| - |Z| + m \leq 2m|Z| - |Z| + |Z| = 2m|Z| \leq 2|Z|^2$$

taking into account that $m \leq |Z|$. For the general case of a rational matrix A, first observe that

$$A = \frac{1}{\prod_{i,j} \underline{a}_{ij}}\ Z,$$

where Z is an integer matrix of the same dimension with each

$$z_{ij} = a_{ij} \prod_{i,j} \underline{a}_{ij}$$

(see Notation 7.11). Clearly,

$$|Z| \leq |A|^2$$

and, so,

$$|\det\ Z| \leq 2|A|^4.$$

Moreover, $|\det\ A| \leq |\prod_{i,j} \underline{a}_{ij}| + |\det\ Z| \leq |A| + 2|A|^4 \leq 3|A|^4.$ QED

We now provide a bound on the size of an optimizer (with rational components) for a standard optimization problem in D^{SF} in terms of the size of the problem.

Proposition 7.5

For every $P = (A, b, c) \in \mathrm{D}^{\mathrm{SF}}$ such that $S_P \neq \emptyset$, there is $s \in S_P$ with rational components such that

$$|s| \leq 9|(A, b)|^5.$$

Proof:

Recall from the proof of Proposition 7.1 that among the rational optimizers of P there is a basic vector s with each component given by

$$s_j = \frac{\det\ (A_B)_j^b}{\det\ A_B}.$$

Therefore, by Proposition 7.4,

$$|s_j| \leq 3|(A_B)^b_i|^4 + 3|A_B|^4 \leq 6|(A,b)|^4.$$

Finally, since $n \leq |(A,b)|$,

$$|s| \leq 6|(A,b)|^4 n + 3n$$
$$\leq 9|(A,b)|^5$$

as required. QED

The previous result is extended to general linear optimization problems (recall Notation 1.6).

Proposition 7.6
Let $P = (A', A'', A''', b', b'', b'', c)$ be a linear optimization problem with rational coefficients in all components such that $S_P \neq \emptyset$. Then, there is $s \in S_P$ with rational components such that

$$|s| \leq d(|A'| + |A''| + |A'''| + |b'| + |b''| + |b'''|)^5$$

for some $d \in \mathbb{N}^+$.

Proof:

Assume that A' is an $m' \times n$-matrix, A'' is an $m'' \times n$-matrix and A''' is an $m''' \times n$-matrix. Observe that $CS(LC(P))$ is the standard optimization problem

$$\begin{cases} \min_y \begin{bmatrix} -c & c & 0 \end{bmatrix} y \\ Ay = b \\ y \geq 0, \end{cases}$$

where A is the $(m' + m'' + 2m''') \times (2n + m' + m'' + 2m''')$-matrix

$$\begin{bmatrix} A' & -A' & I_{m'} & 0 & 0 & 0 \\ -A'' & A'' & 0 & I_{m''} & 0 & 0 \\ A''' & -A''' & 0 & 0 & I_{m'''} & 0 \\ -A''' & A''' & 0 & 0 & 0 & I_{m'''} \end{bmatrix}$$

and b is the vector

$$\begin{bmatrix} b' \\ -b'' \\ b''' \\ -b''' \end{bmatrix}.$$

Therefore,

$$S_{CS(LC(P))} \neq \emptyset$$

by Proposition 1.1 and Proposition 1.3. Hence, by Proposition 7.1, there is $s \in S_{CS(LC(P))}$ with rational components such that

$$|s| \leq 9|(A,b)|^5.$$

Thus, by Proposition 1.3,

$$s|_{2n} \in S_{LC(P)}$$

has only rational components and

$$|s|_{2n}| \leq 9|(A,b)|^5.$$

Finally, by Proposition 1.1,

$$s|_n - s|_{2n}^{n+1} \in S_P$$

and has only rational components and

$$|s|_n - s|_{2n}^{n+1}| \leq 9|(A,b)|^5.$$

The thesis follows since

$$|(A,b)| \leq d'|A'| + d''|A''| + d'''|A'''| + e'|b'| + e''|b''| + e'''|b'''| + e$$

for some $d', d'', d''', e', e'', e''', e > 0$. QED

7.3 Non-Deterministic Approach

We now analyze the complexity of the standard optimization decision problem SDP (see Definition 7.1) and show that it is non-deterministically polynomial (see Definition 7.12).

Proposition 7.7
The standard optimization decision problem SDP is in $\mathcal{NP}$.

Proof:
Take $\mathcal{W}$ to be $\mathbb{Q}^{\cup} \times \mathbb{Q}^{\cup} \times \mathbb{N}^+$ and consider the algorithm $\mathfrak{d}$ in Figure 7.2 with $\mathcal{D}_{\mathfrak{d}} = \mathrm{D}^{\mathrm{SF}} \times (\mathbb{Q}^{\cup} \times \mathbb{Q}^{\cup} \times \mathbb{N}^+)$. Let $P = (A,b,c) \in \mathrm{D}^{\mathrm{SF}}$ and Q be the dual of P. Then,

(1) We show that if $\mathrm{SDP}(P) = 1$ then the execution of $\mathfrak{d}$ on (P,x,y,d) returns

---

input : $((A, b, c), x, y, d)$

1. if $\mathfrak{a}_{\text{sizeGE}}(6\iota(A, b)\iota^5, x)$ $\wedge$ $\mathfrak{a}_{\text{sizeGE}}(d\iota(A, c)\iota^5, y)$ $\wedge$
   $\dim c = \dim x$ $\wedge$ $\dim b = \dim y$ $\wedge$ $x \geq 0$ $\wedge$
   $Ax = b$ $\wedge$ $A^{\mathsf{T}} y \leq c^{\mathsf{T}}$ $\wedge$ $cx = b^{\mathsf{T}} y$ then:

     return 1;

   else:

     return 0.

Figure 7.2: Algorithm $\mathfrak{d}$.

1 for some $(x, y, d) \in \mathbb{Q}^{\cup} \times \mathbb{Q}^{\cup} \times \mathbb{N}^{+}$.
Assume that $\text{SDP}(P) = 1$. Then, $S_P \neq \emptyset$. Hence, by Exercise 6.5,

$$R_Q \neq \emptyset$$

and $cs = b^{\mathsf{T}} r$ for every $s \in S_P$ and $r \in R_Q$. Let x be chosen as in Proposition 7.5 for P and y and d chosen as in Proposition 7.6 with respect to Q. Hence,

$$\iota x \iota \leq 6\iota(A, b)\iota^5$$

and

$$\iota y \iota \leq d(\iota A^{\mathsf{T}} \iota + \iota c^{\mathsf{T}} \iota)^5 \leq d\iota(A, c)\iota^5.$$

Then, the execution of algorithm $\mathfrak{d}$ on (P, x, y, d) returns 1 since the conditions of the if in step 1 are true.

(2) We show that if $\text{SDP}(P) = 0$ then for every $(x, y, d) \in \mathbb{Q}^{\cup} \times \mathbb{Q}^{\cup} \times \mathbb{N}^{+}$ the execution of $\mathfrak{d}$ on (P, x, y, d) returns 0.
Assume that $\text{SDP}(P) = 0$. Thus, $S_P = \emptyset$. Assume, by contradiction, that there is $(x, y, d) \in \mathbb{Q}^{\cup} \times \mathbb{Q}^{\cup} \times \mathbb{N}^{+}$ such that the execution of $\mathfrak{d}$ on (P, x, y, d) returns 1. Then, all the conditions of the if in step 1 are true. Therefore, $x \in X_P$ and $y \in Y_Q$. Moreover, by Exercise 6.4, $x \in S_P$ and $y \in R_Q$, contradicting $S_P = \emptyset$.

(3) Observe that the evaluation of each of the conditions of the if is performed in polynomial time on the size of (A, b, c):

- $\mathfrak{a}_{\text{sizeGE}}(6\iota(A, b)\iota^5, x)$ and $\mathfrak{a}_{\text{sizeGE}}(d\iota(A, c)\iota^5, y)$, by Proposition 7.2;

- $\dim x = \dim c$ and $\dim w = \dim b$, by Proposition 7.11, Proposition 7.3 and the first item;

input : (A, b, c)

1. $s := 0$;

2. $o := +\infty$;

3. for each subset B of $\{1, \dots, n\}$ with m elements do:

    (a) if there is a (unique) solution z of $A_B z = b$ and $z > 0$ then:

    - set x such that
    $$x_j := \begin{cases} z_j & \text{if } j \in B \\ 0 & \text{otherwise}; \end{cases}$$

    - if $cx < o$ then:
        $o := cx$;
        $s := x$;

4. return s.

Figure 7.3: Basic algorithm $\mathfrak{b}$.

- the other conditions, by Proposition 7.12, Exercise 7.4 and the first item.

Thus, $\mathfrak{d}$ is polynomial on the first part of the input; that is, on D^{SF}.   QED

7.4   Deterministic Approach

In this section, we analyze the basic algorithm mentioned in Section 4.2, that relies on a *brute-force* method that goes through all basic admissible vectors, to get an optimizer whenever there is one (recall Theorem 4.1).

Notation 7.5
We say that $\mathfrak{b}$ in Figure 7.3 is the *basic algorithm*.

Proposition 7.8
Let $P = (A, b, c) \in \mathrm{D}^{\mathrm{SF}}$ be a standard optimization problem such that $S_P \neq \emptyset$. Then, the basic algorithm when executed on P returns a vector in S_P.

Proof:
The algorithm is correct since it finds a basic admissible vector which is an

optimizer (see Theorem 4.1). QED

Although $\mathfrak{b}$ is correct, it is, nevertheless, very inefficient due to the high number of times that the body of the cycle in Step 3 is executed. Indeed, observe that the body of the cycle at Step 3 is executed

$$\binom{n}{m} = \frac{n!}{m!(n-m)!}$$

times assuming that A is an $m \times n$-matrix. This fact allows us to show that $\mathfrak{b}$ is not polynomial. To this end, we have to show that there is no polynomial map $f : \mathbb{N} \to \mathbb{R}_0^+$ such that $\omega_{\mathfrak{b}} \in \mathcal{O}(f)$. That is, for every such f we have to provide an unbounded subsequence of the sequence

$$\rho = \nu \mapsto \frac{\omega_{\mathfrak{b}}(\nu)}{f(\nu)}.$$

Definition 7.2
Let $\mu : \mathbb{N} \to \mathbb{N}$ be the map defined as follows:

$$\mu(\nu) = \iota(A, b, c)\iota,$$

where

- $(A, b, c) \in \mathrm{D}^{\mathrm{SF}}$;

- A is a $\frac{n}{2} \times n$-matrix and n is the $(\nu + 1)$-th positive even number;

- each rational entry of A, b and c has 13 bits.

Note that 13 bits is the minimum size of a rational number (see Remark 7.3).

Notation 7.6
We denote by

$$P_\nu$$

the optimization problem corresponding to $\mu(\nu)$ for every $\nu \in \mathbb{N}$. Moreover, we denote by

$$n_\nu$$

the number of columns of the matrix in P_ν.

Proposition 7.9
For each $\nu \in \mathbb{N}$,

$$n_\nu = \frac{\sqrt{2}\sqrt{\mu(\nu) + 21}}{4} - \frac{3}{2}.$$

Proof:

Observe that

- $|c| = 16n_\nu - 3$;

- $|b| = 8n_\nu - 3$;

- $|A| = 8n_\nu^2 - 3$.

For each $\nu \in \mathbb{N}$,

$$|P_\nu| = \mu(\nu) = |(A, b, c)| = |A| + |b| + |c| + 6 = 8n_\nu^2 + 24n_\nu - 3.$$

Observe that, the positive solution of the equation

$$\mu(\nu) = 8n_\nu^2 + 24n_\nu - 3$$

is

$$\frac{\sqrt{2}\sqrt{\mu(\nu) + 21}}{4} - \frac{3}{2}$$

for every $\nu \in \mathbb{N}$.   QED

Proposition 7.10

The basic algorithm is not efficient.

Proof:

Observe that (see [19])

$$\binom{n_\nu}{\frac{n_\nu}{2}} \geq 2^{\frac{n_\nu}{2}}$$

since n_ν is a positive even number. Hence, by Proposition 7.9, we obtain:

$$\binom{n_\nu}{\frac{n_\nu}{2}} \geq 2^{\frac{\sqrt{2}\sqrt{\mu(\nu) + 21}}{8} - \frac{3}{4}}.$$

Therefore, since the cycle in step 3 is executed

$$\binom{n_\nu}{\frac{n_\nu}{2}}$$

times when $\mathfrak{b}$ is run on input P_ν, we get

$$\tau_{\mathfrak{b}}(P_\nu) \geq 2^{\frac{\sqrt{2}\sqrt{\mu(\nu) + 21}}{8} - \frac{3}{4}}$$

and, so,

$$\omega_{\mathfrak{b}}(\mu(\nu)) \geq \tau_{\mathfrak{b}}(P_\nu) \geq 2^{\dfrac{\sqrt{2}\sqrt{\mu(\nu)+21}}{8} - \dfrac{3}{4}}.$$

Hence, $\lim_{\nu \to \infty} \rho(\mu(\nu)) = \lim_{\nu \to \infty} \dfrac{\omega_{\mathfrak{b}}(\mu(\nu))}{f(\mu(\nu))} = +\infty.$ \hfill QED

7.5 Solved Problems and Exercises

---

input : (m, n)

1. while scanning $\hat{m}$ and $\hat{n}$ simultaneously until both end:

   - if the scan of $\hat{m}$ ends before that of $\hat{n}$ then return 1;
   - if the scan of $\hat{n}$ ends before that of $\hat{m}$ then return 0;

2. $o := 1$;

3. while scanning $\hat{m}$ and $\hat{n}$ simultaneously:

   - if the current bit of $\hat{m}$ is greater than the current bit of $\hat{n}$ then $o := 0$;

   - if the current bit of $\hat{m}$ is less than the current bit of $\hat{n}$ then $o := 1$;

4. return o.

Figure 7.4: An algorithm for $\leq^{\mathbb{N}}$.

Problem 7.1
Show that $\leq^{\mathbb{N}} \in \mathcal{P}$.

Solution:
Given $m, n \in \mathbb{N}$. Consider the algorithm in Figure 7.4 that returns the result in o. The first loop performs at most

$$2 \min\{|m|, |n|\} + 3$$

operations. On the other hand, the second loop performs at most

$$6 \min\{|m|, |n|\}$$

---

input : (m, n)
1. $b := 0$;
2. $o := 0$;
3. while scanning $\hat{m}$ and $\hat{n}$ simultaneously until both end:
  - if the scan of $\hat{m}$ has ended then
        $x := 0$;
     else:
           set x to the current bit of $\hat{m}$;
  - if the scan of $\hat{n}$ has ended then:
        $y := 0$;
     else:
           set y to the current bit of $\hat{n}$;
  - set the current bit of $\hat{o}$ to the xor of b, x and y;
  - if there are at least two 1's among b, x, y then:
        $b := 1$;
     else:
        $b := 0$;
4. if $b \neq 0$ then $\hat{o} := 1\hat{o}$;
5. return o;

Figure 7.5: An algorithm for $+^{\mathbb{N}}$

operations. So, the total number of operations is

$$8 \min\{|m|, |n|\} + 5.$$

Therefore, the algorithm is linear and so $\leq^{\mathbb{N}} \in \mathcal{P}$.   ◁

Problem 7.2
Show that $+^{\mathbb{N}} \in \mathcal{P}$.

Solution:
Given $m, n \in \mathbb{N}$, consider the algorithm in Figure 7.5 that returns the result in o and uses b for the carry-over bit. The algorithm computes $m +^{\mathbb{N}} n$ in at most

$$12 \max\{|m|, |n|\} + 5$$

bit operations. Therefore, the algorithm is linear and so $+^{\mathbb{N}} \in \mathcal{P}$.   ◁

Problem 7.3
Show that $\times^{\mathbb{N}} \in \mathcal{P}$.

---

input : (m, n)
1. if $m = 0 \ \vee n = 0$ then return 0;
2. $o := 0$;
3. while scanning $\hat{n}$:
   - if the current bit of $\hat{n}$ is 1 then $o := o +^{\mathbb{N}} m$;
   - $\hat{m} := \hat{m}0$;
4. return o.

Figure 7.6: An algorithm for $\times^{\mathbb{N}}$.

Solution:

Given $m, n \in \mathbb{N}$, consider the algorithm in Figure 7.6. In short, the algorithm notices that,

$$m \times n = m \times \left(\sum_{i \in I} 2^i \right) = \sum_{i \in I} m \times 2^i,$$

where

$$I = \{i : 0 \leq i < \imath n \imath \text{ and } \hat{n}_{\imath n \imath - i} = 1\}.$$

Moreover, recall that multiplying m by 2^i in binary is equivalent to appending to $\hat{m}$ a sequence of size i composed only by 0's. The multiplication algorithm calls the algorithm for $+^{\mathbb{N}}$, and therefore, it is relevant to give a description on how this algorithm is called, to understand its complexity. The main idea is to copy the inputs of the algorithm to some free area of the computer memory, and there, perform the $+^{\mathbb{N}}$ algorithm. Finally, we copy the output back to the program being run.[2] It is easy to see that the loop in step 3 runs for $\imath n \imath$ times. Moreover, the summation in the loop adds numbers with at most $\imath n \imath + \imath m \imath$ bits. Therefore, the number of bit operations in the execution of the algorithm is at most $\imath n \imath (\imath n \imath + \imath m \imath)$ modulo a multiplicative factor. Hence, the worst-case execution time of the algorithm is in $\mathcal{O}(\nu \mapsto \nu^2)$. Thus, it is quadratic and so $\times^{\mathbb{N}} \in \mathcal{P}$. ◁

Problem 7.4
Show that $\leq \, \in \mathcal{P}$.

Solution:
Given $q, r \in \mathbb{Q}$, consider the algorithm in Figure 7.7 that returns the required

---

[2]We stress that there are more efficient multiplication algorithms than the one presented here, namely those based on the Fast Fourier Transform (see for instance [19]).

---

input : (q, r)
1. if $\sigma_q = -\; \wedge\; \sigma_r = +$ then return 1;
2. if $\sigma_q = +\; \wedge\; \sigma_r = -$ then return $(0 \leq^{\mathbb{N}} \overline{q}) \wedge (\overline{q} \leq^{\mathbb{N}} \overline{r}) \wedge (\overline{r} \leq^{\mathbb{N}} 0)$;
3. if $\sigma_q = \sigma_r = +$ then return $\overline{q} \times^{\mathbb{N}} \underline{r} \leq^{\mathbb{N}} \underline{q} \times^{\mathbb{N}} \overline{r}$;
4. if $\sigma_q = \sigma_r = -$ then return $\underline{q} \times^{\mathbb{N}} \overline{r} \leq^{\mathbb{N}} \overline{q} \times^{\mathbb{N}} \underline{r}$.

Figure 7.7: An algorithm for $\leq$.

result using the algorithms for $\leq^{\mathbb{N}}$ and $\times^{\mathbb{N}}$ (see Notation 7.10). The execution times of the algorithms implementing the operations are polynomial on the size of the input. Therefore, the algorithm is efficient. So, $\leq\, \in \mathcal{P}$.   ◁

---

input : (q, r)
1. if $\sigma_q = \sigma_r$ then:
$$\sigma_o := +;$$
    else:
$$\sigma_o := -;$$
2. $(\overline{o}, \underline{o}) := \mathsf{red}(\overline{q} \times^{\mathbb{N}} \overline{r}, \underline{q} \times^{\mathbb{N}} \underline{r})$;
3. return $(\sigma_o, \overline{o}, \underline{o})$.

Figure 7.8: An algorithm for $\times$.

Problem 7.5
Show that $\times\, \in \mathcal{P}$.

Solution:
Consider the algorithm in Figure 7.8 using the algorithms for $\times^{\mathbb{N}}$ and red (see Notation 7.10). Clearly, the algorithm is polynomial. So, $\times\, \in \mathcal{P}$.   ◁

Problem 7.6
Show that $+\, \in \mathcal{P}$.

Solution:
Given $q, r \in \mathbb{Q}$, consider the algorithm in Figure 7.9 that returns the required

---

input : (q, r)
1. if $\sigma_q = \sigma_r$ then:
   - $\sigma_o := \sigma_q$;
   - $(\overline{o}, \underline{o}) := \mathsf{red}(\overline{q} \times^{\mathbb{N}} \underline{r} +^{\mathbb{N}} \underline{q} \times^{\mathbb{N}} \overline{r}, \underline{q} \times^{\mathbb{N}} \underline{r})$;
2. if $\sigma_q = + \ \wedge \ \sigma_r = -$ then:
   - if $\underline{q} \times^{\mathbb{N}} \overline{r} \leq^{\mathbb{N}} \overline{q} \times^{\mathbb{N}} \underline{r}$ then:
     $\sigma_o := +$;
     else:
     $\sigma_o := -$;
   - if $\sigma_o = +$ then:
     $(\overline{o}, \underline{o}) := \mathsf{red}(\overline{q} \times^{\mathbb{N}} \underline{r} \dot{-}^{\mathbb{N}} \underline{q} \times^{\mathbb{N}} \overline{r}, \underline{q} \times^{\mathbb{N}} \underline{r})$;
     else:
     $(\overline{o}, \underline{o}) := \mathsf{red}(\underline{q} \times^{\mathbb{N}} \overline{r} \dot{-}^{\mathbb{N}} \overline{q} \times^{\mathbb{N}} \underline{r}, \underline{q} \times^{\mathbb{N}} \underline{r})$;
3. if $\sigma_q = - \ \wedge \ \sigma_r = +$ then:
   - if $\overline{q} \times^{\mathbb{N}} \underline{r} \leq^{\mathbb{N}} \underline{q} \times^{\mathbb{N}} \overline{r}$ then:
     $\sigma_o := +$;
     else:
     $\sigma_o := -$;
   - if $\sigma_o = -$ then:
     $(\overline{o}, \underline{o}) := \mathsf{red}(\overline{q} \times^{\mathbb{N}} \underline{r} \dot{-}^{\mathbb{N}} \underline{q} \times^{\mathbb{N}} \overline{r}, \underline{q} \times^{\mathbb{N}} \underline{r})$;
     else:
     $(\overline{o}, \underline{o}) := \mathsf{red}(\underline{q} \times^{\mathbb{N}} \overline{r} \dot{-}^{\mathbb{N}} \overline{q} \times^{\mathbb{N}} \underline{r}, \underline{q} \times^{\mathbb{N}} \underline{r})$;
4. return $(\sigma_o, \overline{o}, \underline{o})$.

Figure 7.9: An algorithm for $+$.

result $(\sigma_o, \overline{o}, \underline{o})$ using the algorithms for $\leq^{\mathbb{N}}$, $+^{\mathbb{N}}$, $\dot{-}^{\mathbb{N}}$, $\times^{\mathbb{N}}$ and red (see Notation 7.10). It is immediate that the algorithm is polynomial on the size of the input. So, $+ \in \mathcal{P}$. ◁

Problem 7.7

Consider the decision problem

$$\mathrm{Prime} : \mathbb{N} \to \{0, 1\}$$

such that $\mathrm{Prime}(x) = 1$ if and only if x is a prime number. Show that Prime is in $\mathcal{P}$ (see Definition 7.9).

Solution:

There are several algorithms, usually called *primality tests* for this problem based on different characterization of a prime number. The Wilson Theorem (see [39]) states that given $p \in \mathbb{N}$ with $p \neq 1$

$$p \text{ is a prime number} \quad \text{iff} \quad (p-1)! \equiv -1 \ (\text{mod } p).$$

For instance, 5 is a prime number since $4! + 1 = 24 + 1$ is a multiple of 5. There is a straightforward algorithm based on this result. However, such an algorithm is not polynomial. On the other hand, a generalization to polynomials of the Fermat's Little Theorem (see [39]) states that for $p \geq 2$,

$$p \text{ is prime} \quad \text{iff} \quad \text{there is } q \text{ coprime } p \text{ with } (x+q)^p \equiv (x^p + q) \ (\text{mod } p).$$

For instance, let $p = 2$ and $q = 1$. Then, $2x$ is always a multiple of 2. Hence, 2 is a prime number.

It was shown in [1] that there is an efficient algorithm, that we refer to as $\mathfrak{a}_{\text{Prime}}$, based on this result for Prime. Therefore, Prime is in $\mathcal{P}$.   ◁

Problem 7.8
Consider the decision problem

$$\text{Factor} : \mathbb{N}^3 \rightarrow \{0, 1\}$$

such that $\text{Factor}(d, k_1, k_2) = 1$ if and only if d has a prime factor in $\{k_1, \ldots, k_2\}$. For instance, $\text{Factor}(21, 2, 5) = 1$ and $\text{Factor}(21, 10, 11) = 0$. Show that

$$\text{Factor} \in \mathcal{NP}$$

(see Definition 7.12).

Solution:
Let

$$\mathcal{W} = \mathbb{N}$$

and

$$\mathfrak{a}_{\text{Fact}} : \mathbb{N}^3 \times \mathcal{W} \rightarrow \{0, 1\}$$

be the algorithm in Figure 7.10, using the efficient algorithm $\mathfrak{a}_{\text{Prime}}$ mentioned in Problem 7.7. Assume that

$$\text{Factor}(d, k_1, k_2) = 1.$$

So, there is a prime number p such that $k_1 \leq p \leq k_2$ and d is a multiple of p. Take p as the witness. Then, the execution of algorithm $\mathfrak{a}_{\text{Fact}}$ on the input $((d, k_1, k_2), p)$ terminates with value 1. Indeed:

---

input : $((d, k_1, k_2), p)$

1. if $p > k_2 \ \vee \ p < k_1 \ \vee \ \mathfrak{a}_{\mathrm{Prime}}(p) = 0$ then:

   return 0;

   else:

   (a) if $\mathrm{mod}(d, p) = 0$ then:

   return 1;

   else:

   return 0.

Figure 7.10: Algorithm $\mathfrak{a}_{\mathrm{Fact}}$.

- conditions $p > k_2$, $p < k_1$ and $\alpha_{\mathrm{Prime}}(p) = 0$ are false;
- condition $\mathrm{mod}(d, p) = 0$ is true;
- the output is 1.

Assume that
$$\mathrm{Factor}(d, k_1, k_2) = 0.$$
So, there no prime number p such that $k_1 \leq p \leq k_2$ and d is a multiple of p. Take $q \in \mathbb{N}$ as an arbitrary witness. Then, the execution of algorithm $\mathfrak{a}_{\mathrm{Fact}}$ on the input $((d, k_1, k_2), q)$ terminates with value 0. Indeed:

- either $q > k_2$ or $q < k_1$ or $\mathfrak{a}_{\mathrm{Prime}}(q) = 0$;
- for any of the cases above the output is 0.

Since $\mathfrak{a}_{\mathrm{Prime}}$ is efficient, it is very easy to show that $\mathfrak{a}_{\mathrm{Fact}}$ is polynomial on $\mathbb{N}^3$. Thus, Factor is a $\mathcal{NP}$ problem. ◁

Problem 7.9
Consider the maps $f = \nu \mapsto a_n \nu^n + \cdots + a_0$ and $g = \nu \mapsto \nu^n$. Show that
$$f \in \mathcal{O}(g).$$

Solution:
We have
$$a_n \nu^n + \cdots + a_0 \leq \nu^n (|a_n| + \cdots + |a_0|).$$
The thesis follows taking $\alpha = |a_n| + \cdots + |a_0|$ and $\nu_0 = 0$. ◁

Exercise 7.1

Show that $=^{\mathbb{N}}, \div^{\mathbb{N}}, \mathsf{rem}, \mathsf{gcd}, \mathsf{red} \in \mathcal{P}$ (see Notation 7.10).

Exercise 7.2

Show that $\bar{\div}, \underline{\cdot}, \sigma., \mathsf{I} \cdot \mathsf{I}, =, -, \cdot^{-1} \in \mathcal{P}$ (see Notation 7.12).

Exercise 7.3

Show that there exists a polynomial algorithm for each of the following operations:

- $w \mapsto \mathsf{I}w\mathsf{I} : \mathbb{Q}^{\cup} \to \mathbb{N}^{+}$;

- $A \mapsto \mathsf{I}A\mathsf{I} : \mathbb{Q}^{\cup} \times \mathbb{Q}^{\cup} \to \mathbb{N}^{+}$;

- $(A, b, c) \mapsto \mathsf{I}(A, b, c)\mathsf{I} : \mathrm{D}^{\mathrm{SF}} \to \mathbb{N}^{+}$;

- $(A, b) \mapsto \mathsf{I}(A, b)\mathsf{I} : \{(A, b) : \exists c \; (A, b, c) \in \mathrm{D}^{\mathrm{SF}}\} \to \mathbb{N}^{+}$;

- $(A, c) \mapsto \mathsf{I}(A, c)\mathsf{I} : \{(A, c) : \exists b \; (A, b, c) \in \mathrm{D}^{\mathrm{SF}}\} \to \mathbb{N}^{+}$.

Exercise 7.4

Show that there is a polynomial algorithm for computing each of the following linear algebra operations over the field $\mathbb{Q}$:

- inner product of two vectors;

- transpose of a matrix;

- product of a matrix by a vector;

- entry-wise comparisons ($\leq$ and $=$) between vectors.

Exercise 7.5

Let $\mathrm{CVP} : \mathrm{D}^{\mathrm{SF}} \times \mathbb{Q}^{\cup} \to \{0, 1\}$ be a map such that

$$\mathrm{CVP}(P, x) = 1 \quad \text{if and only if} \quad x \in S_P$$

for every pure canonical optimization problem P. Use the active lines technique (see Section 3.2) to conclude that $\mathrm{CVP} \in \mathcal{P}$.

Exercise 7.6

Let $g : \mathbb{N} \to \mathbb{R}_0^+$. Show that

$$\mathcal{O}(g) = \left\{ f : \mathbb{N} \to \mathbb{R}_0^+ \ : \ \limsup_{\nu \to \infty} \frac{f(\nu)}{g(\nu)} < \infty \right\}.$$

Exercise 7.7

Let $f, g, h : \mathbb{N} \to \mathbb{R}_0^+$ be maps. Show that

- $\mathcal{O}(g + h) = \mathcal{O}(\max\{g, h\})$;
- $\mathcal{O}(\beta g) = \mathcal{O}(g)$, for each $\beta \in \mathbb{R}^+$;
- If $f_j \in \mathcal{O}(g_j)$ for $j = 1, 2$ then $f_1 + f_2 \in \mathcal{O}(g_1 + g_2)$;
- $\mathcal{O}(g)$ is closed for the sum of maps;
- $f + \mathcal{O}(g) = \{f + h : h \in \mathcal{O}(g)\} \subseteq \mathcal{O}(f + g)$;
- $\mathcal{O}(g) \cup \mathcal{O}(h) \subseteq \mathcal{O}(\max\{g, h\})$;
- $\mathcal{O}(\min\{g, h\}) \subseteq \mathcal{O}(g) \cup \mathcal{O}(h)$;

where

- βf for $\nu \mapsto \beta f(\nu)$;
- $f + g$ for $\nu \mapsto f(\nu) + g(\nu)$;
- $\max\{f, g\}$ for $\nu \mapsto \max\{f(\nu), g(\nu)\}$;
- $\min\{f, g\}$ for $\nu \mapsto \min\{f(\nu), g(\nu)\}$.

7.6 Relevant Background

To analyze the efficiency of algorithms for solving linear optimization problems, we need some concepts and results from the theory of computational complexity [34, 5]. The efficiency of an algorithm is measured by the computational resources that are required for its execution as a map of the size of the input data. These resources are the execution time and the memory space.

Notation 7.7

Given an algorithm $\mathfrak{a}$, we denote by

$$\mathcal{D}_\mathfrak{a}$$

the *input data space*.

Remark 7.1
In the sequel, we assume that $\mathcal{D}_{\mathfrak{a}}$ is a denumerable set.

Notation 7.8
For each $d \in \mathcal{D}_{\mathfrak{a}}$, we denote by

$$|d|$$

the *size* of d; that is, the number of bits used to represent d.

Notation 7.9
Given an algorithm $\mathfrak{a}$, we denote by

$$\tau_{\mathfrak{a}} : \mathcal{D}_{\mathfrak{a}} \to \mathbb{R}_0^+$$

the map assigning to each d the *execution time* of $\mathfrak{a}$ on the input d; that is, the number of bit operations that $\mathfrak{a}$ executes on d.

Definition 7.3
The *worst-case execution time* of an algorithm $\mathfrak{a}$ is the map

$$\omega_{\mathfrak{a}} = \nu \mapsto \max\{\tau_{\mathfrak{a}}(d) : d \in \mathcal{D}_{\mathfrak{a}}, |d| \leq \nu\} : \mathbb{N} \to \mathbb{R}_0^+.$$

In general, the idea is not to obtain the map $\omega_{\mathfrak{a}}$ explicitly but rather to identify a class of maps where this map belongs. In particular, from an efficiency point of view, it is desirable that $\omega_{\mathfrak{a}}$ belongs to a polynomial class.

For this purpose, we need to define the *big $\mathcal{O}$ notation* (also called the Landau or the Bachman-Landau notation and in some cases also called the asymptotic notation, see [19]).

Definition 7.4
The *class of maps with asymptotic upper bound $g : \mathbb{N} \to \mathbb{R}_0^+$ up to a multiplicative constant*, denoted by

$$\mathcal{O}(g),$$

is the set of maps $f : \mathbb{N} \to \mathbb{R}_0^+$ such that

$$\exists \alpha \in \mathbb{R}^+ \, \exists \nu_0 \in \mathbb{N} \, \forall \nu > \nu_0 \ \ f(\nu) \leq \alpha g(\nu).$$

Definition 7.5
An algorithm $\mathfrak{a}$ is *polynomial* or *algebraic* or *efficient* (in time), written

$$\omega_\mathfrak{a} \in \mathcal{P} \quad \text{or even} \quad \mathfrak{a} \in \mathcal{P}$$

if

$$\exists\, k \in \mathbb{N} \quad \omega_\mathfrak{a} \in \mathcal{O}(\nu \mapsto \nu^k)$$

(see Remark 1.1). When $\mathfrak{a}$ is polynomial, we denote by

$$k_\mathfrak{a}$$

the minimum value of k satisfying the condition.

Definition 7.6
We say that a polynomial algorithm $\mathfrak{a}$ is *linear* when $k_\mathfrak{a} = 1$ and say that it is *quadratic* when $k_\mathfrak{a} = 2$.

Definition 7.7
An algorithm $\mathfrak{a}$ is *exponential* if

$$\exists\, k \in \mathbb{N} \quad \omega_\mathfrak{a} \in \mathcal{O}(\nu \mapsto 2^{\nu^k}).$$

Definition 7.8
A *computation problem* is a map $h : \mathcal{D} \to \mathcal{E}$ where $\mathcal{D}$ is a denumerable set and $\mathcal{E}$ is a countable set. A *decision problem* is a computation problem where $\mathcal{E} = \{0, 1\}$.[3]

Definition 7.9
A computation problem $h : \mathcal{D} \to \mathcal{E}$ is *polynomial*, written

$$h \in \mathcal{P},$$

if there exists an efficient algorithm $\mathfrak{a}$ such that:

- $\mathcal{D}_\mathfrak{a} = \mathcal{D}$;

- for each $d \in \mathcal{D}$, the execution of $\mathfrak{a}$ on input d returns $h(d)$.

---

[3]The terminology of decision problem was first introduced by the mathematician David Hilbert in 1928 (see [37]) when asking whether or not Mathematics is decidable.

Sometimes, a problem is difficult to solve but checking if a candidate is indeed a solution of the problem is much easier. This intuition leads to another relevant class of complexity composed of the problems whose solutions can be verified in polynomial time.

Before defining the class of such problems, we need to introduce the notion of an algorithm being *polynomial only on part of the input.*

Definition 7.10
Let $\mathfrak{a}$ be an algorithm with input data space factorized as follows:

$$\mathcal{D}_{\mathfrak{a}} = \mathcal{D}'_{\mathfrak{a}} \times \mathcal{D}''_{\mathfrak{a}}.$$

The *worst-case execution time based on the first argument* of $\mathfrak{a}$, denoted by

$$\omega^1_{\mathfrak{a}},$$

is the map

$$\nu \mapsto \sup\{\tau_{\mathfrak{a}}(d', d'') : (d', d'') \in \mathcal{D}_{\mathfrak{a}}, |d'| \leq \nu\} : \mathbb{N} \to \overline{\mathbb{R}}^+_0.$$

Definition 7.11
An algorithm $\mathfrak{a}$ with $\mathcal{D}_{\mathfrak{a}} = \mathcal{D}'_{\mathfrak{a}} \times \mathcal{D}''_{\mathfrak{a}}$ is *polynomial on* $\mathcal{D}'_{\mathfrak{a}}$ if

$$\exists k \in \mathbb{N} \quad \omega^1_{\mathfrak{a}} \in \mathcal{O}(\nu \mapsto \nu^k).$$

Definition 7.12
A decision problem $h : \mathcal{D} \to \{0, 1\}$ is *non-deterministically polynomial*, written

$$h \in \mathcal{NP},$$

if there exist a set $\mathcal{W}$ and an algorithm $\mathfrak{a}$ with $\mathcal{D}_{\mathfrak{a}} = \mathcal{D} \times \mathcal{W}$ polynomial on $\mathcal{D}$ and such that:

- if $h(d) = 1$ then there exists $w \in \mathcal{W}$ such that the execution of $\mathfrak{a}$ on input (d, w) returns 1;

- if $h(d) = 0$ then, for every $w \in \mathcal{W}$, the execution of $\mathfrak{a}$ on input (d, w) returns 0.

In this situation, the elements of $\mathcal{W}$ are called *witnesses* or *certificates*. Clearly, every decision problem $h : \mathcal{D} \to \{0, 1\}$ in $\mathcal{P}$ is in $\mathcal{NP}$. The converse is still an open problem,[4] although the majority of mathematicians believe in the conjecture[5]

$$\mathcal{P} \neq \mathcal{NP}.$$

To assess the efficiency of algorithms and establish the complexity of problems within the realm of rational numbers, we need first to analyze the time complexity of the basic operations (including addition, multiplication and comparison) on rational numbers represented with infinite precision by pairs of coprime natural numbers.

Moreover, we assume a *binary representation* for each natural number n, denoted by

$$\hat{n} = \hat{n}_1 \ldots \hat{n}_{(n)},$$

where each $\hat{n}_i \in \{0, 1\}$, (n) is the *length* of the binary representation of n, and $\hat{n}_1$ is its *most significant bit*. Recall that either $\hat{n}_1 = 1$ or $n = 0$. In the latter case $(n) = 1$. From this point on, we assume that the representation of n is $\hat{n}$ and, so, that $|n| = (n)$.

We assume also that there is independent and direct access to the least significant bit of each of the inputs of an operation and that afterwards each input can be sequentially scanned from bit to bit until the most significant bit is reached.

We start by analyzing the time complexity of the following relevant basic operations on natural numbers.

Notation 7.10
The following operations on natural numbers are relevant in the chapter:

$\leq^{\mathbb{N}} : m, n \mapsto m$ less than or equal to $n : \mathbb{N}^2 \to \{0, 1\}$;

$=^{\mathbb{N}} : m, n \mapsto m$ equal to $n : \mathbb{N}^2 \to \{0, 1\}$;

$+^{\mathbb{N}} : m, n \mapsto$ sum of m and $n : \mathbb{N}^2 \to \mathbb{N}$;

$\overset{\cdot}{-}^{\mathbb{N}} : m, n \mapsto$ maximum between 0 and $n - m : \mathbb{N}^2 \to \mathbb{N}$;

$\times^{\mathbb{N}} : m, n \mapsto$ product of m and $n : \mathbb{N}^2 \to \mathbb{N}$;

$\div^{\mathbb{N}} : m, n \mapsto$ integer division of m by $n : \mathbb{N}^2 \to \mathbb{N}$;

$\mathsf{rem} : m, n \mapsto$ reminder of division of m by $n : \mathbb{N}^2 \to \mathbb{N}$;

$\mathsf{gcd} : m, n \mapsto$ greatest common divisor of m and $n : \mathbb{N}^2 \to \mathbb{N}$;

$\mathsf{red} : m, n \mapsto (m \div^{\mathbb{N}} \mathsf{gcd}(m, n), n \div^{\mathbb{N}} \mathsf{gcd}(m, n)) : \mathbb{N} \times \mathbb{N}^+ \to \mathbb{N} \times \mathbb{N}^+$.

---

[4]Problem 4 of the *Millenium Problems* from the Clay Mathematics Institute.
[5]Those interested in the problem should read Stephen Cook's article [18] for a very nice presentation of the main issues related to this problem.

The reader will expect that adding two natural numbers with a few bits will take less time than making the same operation on two natural numbers with many bits. The key assumption here is that *each single-bit operation, including scanning a bit, comparing bits, summing two bits, assigning a bit to a variable, among others, takes a unit of time.*

Proposition 7.11

There exists a polynomial algorithm for each operation in Notation 7.10.

The proof of the previous proposition is spread across Problem 7.1, Problem 7.2, Problem 7.3 and Exercise 7.1.

Actually, for showing that all these operations can be computed in polynomial time, we need only sufficiently good upper bounds on the time complexity of these operations (see Problem 7.1, Problem 7.2 and Problem 7.3). Nevertheless, when available, we mention *en passant* better upper bounds that one can easily find in the literature but that are more difficult to establish.

Capitalizing on the efficient algorithms presented in the proof of Proposition 7.11, it is straightforward to provide efficient algorithms for the basic operations on rational numbers. To this end, we start with the concept of coprime natural numbers.

Definition 7.13

Two natural numbers are *coprime* if their greatest common divisor is 1.

Remark 7.2

Observe that the map

$$(m, n) \mapsto m/n : \{(m, n) \in \mathbb{N} \times \mathbb{N}^+ : m \text{ and } n \text{ are coprime}\} \to \mathbb{Q}_0^+$$

is a bijection since the coprimality requirement guarantees that each non-negative rational number has a unique representation. This representation is extended as expected to $\mathbb{Q}$ by adding a sign bit. Note that this extension is no longer bijective because 0 has two signed representations.

Notation 7.11

Given a non-negative rational number q, we refer to its unique coprimal representation by

$$(\overline{q}, \underline{q}).$$

Moreover, each $q \in \mathbb{Q}$ is identified with the triple

$$(\sigma_q, \overline{q}, \underline{q}).$$

For representing a rational number

$$(\sigma_q, \overline{q}, \underline{q}),$$

in the computer memory, we need to encode the sign σ_q, the natural numbers $\overline{q}$ and $\underline{q}$, and the separators between the natural numbers.

Remark 7.3
The encoding

$$\hat{\sigma}_q$$

of the sign is 1 if σ_q is $+$ and is 0 otherwise. To encode the comma ",", we use three bits, namely 001. Finally, for the representation of natural numbers, we need to distinguish them from the separator. Hence, we triplicate each bit in its binary representation. If

$$\overline{q} = \sum_{i=1}^{|\overline{q}|} 2^{i-1} b_i$$

then, we use the following string

$$\hat{\overline{q}} = b_{|\overline{q}|} b_{|\overline{q}|} b_{|\overline{q}|} \cdots b_2 b_2 b_2 b_1 b_1 b_1$$

to represent $\overline{q}$. Similarly, for $\underline{q}$. In this way, a rational is represented by the string

$$\hat{q} = \hat{\sigma}_q \, 001 \, \hat{\overline{q}} \, 001 \, \hat{\underline{q}}.$$

Observe that this string requires $|q| = 7 + 3|\overline{q}| + 3|\underline{q}|$ bits (including one bit for the sign and six bits for the two separators 001). Thus, the *minimum size of a rational number* is 13 bits.

Notation 7.12
The following operations on rational numbers are relevant in the chapter:

$$\overline{\cdot} : q \mapsto \overline{q} : \mathbb{Q} \to \mathbb{N};$$
$$\underline{\cdot} : q \mapsto \underline{q} : \mathbb{Q} \to \mathbb{N}^+;$$
$$\sigma. : q \mapsto \sigma_q : \mathbb{Q} \to \{+, -\};$$
$$|\cdot| : \mapsto |q| : \mathbb{Q} \to \mathbb{N}^+;$$
$$\leq : q, r \mapsto q \text{ less than or equal to } r : \mathbb{Q}^2 \to \{0, 1\};$$
$$= : q, r \mapsto q \text{ equal to } r : \mathbb{Q}^2 \to \{0, 1\};$$
$$+ : q, r \mapsto \text{ sum of } q \text{ and } r : \mathbb{Q}^2 \to \mathbb{Q};$$
$$\times : q, r \mapsto \text{ product of } q \text{ and } r : \mathbb{Q}^2 \to \mathbb{Q};$$
$$- : q \mapsto \text{ symmetric of } q : \mathbb{Q} \to \mathbb{Q};$$
$$\cdot^{-1} : q \mapsto 1/q : \mathbb{Q}^+ \cup \mathbb{Q}^- \to \mathbb{Q}.$$

The idea is to reduce operations on rational numbers to operations on natural numbers. In this way, we can take advantage on the algorithms introduced above on natural numbers (see Proposition 7.11) to provide algorithms on rational numbers.

Proposition 7.12
There exists a polynomial algorithm for each of the operations in the rational numbers.

The proof of the previous proposition is spread across Problem 7.4, Problem 7.5, Problem 7.6 and Exercise 7.2.

Chapter 8

The Simplex Algorithm

Herein, we present the Simplex Algorithm that provides a (positive or negative) answer to the existence of optimizers of standard optimization problems and provides an optimizer in the positive case (see [22, 24, 25, 49, 2, 32]). We prove the soundness and completeness of the algorithm. Finally, we show that the Simplex Algorithm is not polynomial.

8.1 Algorithm

In this section we discuss the Simplex Algorithm. We assume given a standard optimization problem resulting from a canonical one. This is not a limitation since we can always get a canonical optimization problem from a general linear optimization problem (see Proposition 1.1). When an optimization problem is such that $b \not> 0$, we can use either Exercise 1.4 or Exercise 1.5 to obtain an equivalent optimization problem with a positive restriction vector.

Definition 8.1
Let $P \in CS(\mathcal{C})$ be a non-degenerate problem with $b > 0$. The *Simplex Algorithm* for P is the algorithm in Figure 8.1.

The variables b and s contain the output results. The variable b is equal to 0 when the problem has no optimizers and is 1 otherwise. The variable s returns an optimizer whenever there is one. The variables x, P, y, k, t, o and j are auxiliary. The auxiliary variable x contains at each point the basic admissible vector under analysis. The variable P is the set P_x. The variable y is a potential admissible vector of the dual problem. When y is admissible, then

---

input : (A, b, c) where A is an $m \times (n + m)$-matrix

1. $\mathsf{x} := (0, \ldots, 0, b_1, \ldots, b_m)$;

2. $\mathsf{P} := \{j \in \{1, \ldots, n + m\} : \mathsf{x}_j > 0\}$;

3. $\mathsf{y} :=$ solution of the system $(A_{\mathsf{P}})^{\mathsf{T}} y = (c_{\mathsf{P}})^{\mathsf{T}}$;

4. if $A^{\mathsf{T}}\mathsf{y} \leq c^{\mathsf{T}}$ then:

   (a) $(\mathsf{b}, \mathsf{s}) := (1, \mathsf{x})$;

   (b) return (b, s);

  else:

   (a) $\mathsf{k} :=$ some k such that $(A^{\mathsf{T}}\mathsf{y})_k > (c^{\mathsf{T}})_k$;

   (b) $\mathsf{t} := t$ with $t_{\mathsf{P}} = (A_{\mathsf{P}})^{-1} a_{\bullet \mathsf{k}}$, $t_{\mathsf{k}} = -1$, $t_{\{1, \ldots, n+m\} \setminus (\mathsf{P} \cup \{\mathsf{k}\})} = 0$;

   (c) if $\mathsf{t} \leq 0$, then:

      i. $\mathsf{b} := 0$;

      ii. return (b, x);

    else:

      i. $\mathsf{o} := \min\left\{\dfrac{\mathsf{x}_j}{\mathsf{t}_j} : \mathsf{t}_j > 0\right\}$;

      ii. $\mathsf{x} := \mathsf{x} - \mathsf{o}\mathsf{t}$;

      iii. go to 2.

Figure 8.1: Simplex Algorithm s.

x is an optimizer. Otherwise, we take note in k of a component that witnesses that y is not admissible. The variables t and o are used to change x to another basic admissible vector.

We now illustrate the application of the Simplex Algorithm.

Example 8.1
Recall the standard optimization problem P

$$\begin{cases} \min_{x} -2x_1 - x_2 \\ 3x_1 - x_2 + x_3 = 6 \\ -x_1 + 3x_2 + x_4 = 6 \\ x \geq 0 \end{cases}$$

introduced in Example 1.13. Observe that $P \in CS(C)$ (see Example 1.12). This problem is shown to be non-degenerate in Example 4.5. Morever, $b > 0$. Hence, we can apply algorithm s to P. Let

$$\mathsf{x} = \begin{bmatrix} 0 \\ 0 \\ 6 \\ 6 \end{bmatrix} \qquad \mathsf{P} = \{3, 4\}$$

By solving the system of equations $(A_{\mathsf{P}})^{\mathsf{T}} y = (c_{\mathsf{P}})^{\mathsf{T}}$ we get

$$y = \begin{bmatrix} 0 \\ 0 \end{bmatrix}.$$

The vector y does not satisfy

$$A^{\mathsf{T}} y \leq c^{\mathsf{T}}.$$

since it fails to satisfy both inequalities. Then, either $k = 1$ or $k = 2$. Assume we choose $k = 1$. Then,

$$\mathsf{t} = \begin{bmatrix} -1 \\ 0 \\ 3 \\ -1 \end{bmatrix}.$$

Since it is not the case that $\mathsf{t} \leq 0$, then, $\mathsf{o} = \min \left\{ \dfrac{6}{3} \right\} = 2$. Hence,

$$\mathsf{x} = \begin{bmatrix} 2 \\ 0 \\ 0 \\ 8 \end{bmatrix}.$$

In this case, $\mathsf{P} = \{1,4\}$. By solving the system of equations $(A_\mathsf{P})^\mathsf{T} y = (c_\mathsf{P})^\mathsf{T}$ we get

$$y = \begin{bmatrix} -\dfrac{2}{3} \\ 0 \end{bmatrix}.$$

The vector y does not satisfy

$$A^\mathsf{T} y \leq c^\mathsf{T}$$

since the second inequality does not hold. Thus, $\mathsf{k} = 2$. Then,

$$t = \begin{bmatrix} -\dfrac{1}{3} \\ -1 \\ 0 \\ \dfrac{8}{3} \end{bmatrix}.$$

Since it is not the case that $\mathsf{t} \leq 0$, then, $\mathsf{o} = 3$. Hence,

$$x = \begin{bmatrix} 3 \\ 3 \\ 0 \\ 0 \end{bmatrix}.$$

So, $\mathsf{P} = \{1,2\}$. By solving the system of equations $(A)_M y = (c)_M^\mathsf{T}$ we get

$$y = \begin{bmatrix} -\dfrac{7}{8} \\ -\dfrac{5}{8} \end{bmatrix}$$

Vector y satisfies

$$A^\mathsf{T} y \leq c^\mathsf{T}.$$

Therefore, s outputs $\mathsf{b} = 1$ and $\mathsf{s} = (3,3,0,0)$.

As we have seen in the example above, it may be the case that there are 2 or more inequalities in $A^\mathsf{T} y \leq c^\mathsf{T}$ that do not hold, for a particular y. Therefore, there are several possible choices for k. One of the most well known choices is the *gradient* choice (which was used by George Dantzig in his seminal paper [22]) consisting in choosing $\mathsf{k} = k$ where k is the smallest element in

$$\{j \in N : c_j - (A^\mathsf{T} y)_j < 0\}.$$

The choice of k in Example 8.1 was done according to this criterion.

8.2 Soundness, Completeness and Complexity

The objective of this section is to prove the soundness and completeness of the Simplex Algorithm. We start by showing some auxiliary lemmas.

Proposition 8.1
Let $P = (A, b, c) \in CS(\mathcal{C})$ be a non-degenerate problem with $b > 0$. Then, $(0, \ldots, 0, b_1, \ldots, b_m)$, in Step 1 of the Simplex Algorithm in Figure 8.1, is a basic admissible vector.

Proof:
We start by observing that $(0, \ldots, 0, b_1, \ldots, b_m) \in X_P$, since $b > 0$ by hypothesis. Moreover, $(0, \ldots, 0, b_1, \ldots, b_m)$ is a basic vector by Proposition 4.9, because $b > 0$. QED

Observe that the system of equations in Step 3 of the Simplex Algorithm has always a unique solution. This is the case since A_P is a nonsingular matrix because the problem is non-degenerate and, in each iteration, x is a basic admissible vector (see Proposition 8.5).

Proposition 8.2
Let $P = (A, b, c) \in CS(\mathcal{C})$ be a non-degenerate problem with $b > 0$. When $x \in X_P$ and the guard of Step 4 of the Simplex Algorithm in Figure 8.1 holds, then $x \in S_P$.

Proof:
Let Q be the dual of P (see Exercise 6.2). Since $A^\mathsf{T} y \leq c^\mathsf{T}$, then $y \in Y_Q$. Furthermore, after assignment at Step 3,

$$A_\mathsf{P}{}^\mathsf{T} y = c_\mathsf{P}{}^\mathsf{T}.$$

Hence,

$$x_\mathsf{P}^\mathsf{T} A_\mathsf{P}{}^\mathsf{T} y = x_\mathsf{P}^\mathsf{T} c_\mathsf{P}{}^\mathsf{T}.$$

Thus,

$$(A_\mathsf{P} x_\mathsf{P})^\mathsf{T} y = x_\mathsf{P}^\mathsf{T} c_\mathsf{P}{}^\mathsf{T}.$$

Therefore,

$$(Ax)^\mathsf{T} y = x^\mathsf{T} c^\mathsf{T}.$$

So,

$$b^\mathsf{T} y = cx,$$

since $Ax = b$. Taking into account that $\mathsf{x} \in X_P$ and $\mathsf{y} \in Y_Q$, by applying Proposition 6.4, we conclude that $\mathsf{x} \in S_P$.   QED

Proposition 8.3
Let $P = (A, b, c) \in CS(\mathcal{C})$ be a non-degenerate problem with $b > 0$. When $\mathsf{x} \in X_P$, the guard of Step 4 of the Simplex Algorithm in Figure 8.1 does not hold and $\mathsf{t} \leq 0$ then $S_P = \emptyset$.

Proof:
Consider the family

$$\{w^\delta\}_{\delta \in \mathbb{R}_0^+},$$

where

$$w^\delta = \mathsf{x} - \delta \mathsf{t}.$$

We start by showing that $\mathsf{k} \notin \mathsf{P}$. Indeed,

$$(A^\mathsf{T} \mathsf{y})_\mathsf{k} > (c^\mathsf{T})_\mathsf{k}.$$

On the other hand,

$$(A_\mathsf{P})^\mathsf{T} \mathsf{y} = (c_\mathsf{P})^\mathsf{T}.$$

That is,

$$(A^\mathsf{T} \mathsf{y})_j = (c^\mathsf{T})_j$$

for every $j \in \mathsf{P}$. Thus,

$$\mathsf{k} \notin \mathsf{P}.$$

Then, for every $j \in N$, we have:

$$w_j^\delta = \begin{cases} \delta & \text{if } j = \mathsf{k} \\ \mathsf{x}_j - \delta \mathsf{t}_j & \text{if } j \in \mathsf{P} \\ 0 & \text{otherwise.} \end{cases}$$

We now show that

$$cw^\delta < c\mathsf{x}$$

for every $\delta > 0$. Indeed,

$$
\begin{aligned}
cw^\delta &= c_k\delta + c_\mathsf{P}x_\mathsf{P} - \delta\,c_\mathsf{P}t_\mathsf{P} \\
&= c_\mathsf{P}x_\mathsf{P} + \delta(c_k - c_\mathsf{P}t_\mathsf{P}) \\
&= c_\mathsf{P}x_\mathsf{P} + \delta(c_k - y^\mathsf{T}A_\mathsf{P}\,t_\mathsf{P}) &(*) \\
&= c_\mathsf{P}x_\mathsf{P} + \delta(c_k - y^\mathsf{T}A_\mathsf{P}\,(A_\mathsf{P})^{-1}a_{\bullet k}) &(**) \\
&= c_\mathsf{P}x_\mathsf{P} + \delta(c_k - y^\mathsf{T}a_{\bullet k}) &(\dagger) \\
&< c_\mathsf{P}x_\mathsf{P} &(***) \\
&= cx,
\end{aligned}
$$

where the justifications are as follows:

$(*)$ since $(A_\mathsf{P})^\mathsf{T}y = (c_\mathsf{P})^\mathsf{T}$.

$(**)$ because $t_\mathsf{P} = (A_\mathsf{P})^{-1}a_{\bullet k}$.

$(***)$ since $(A^\mathsf{T}y)_k > (c^\mathsf{T})_k$.

Moreover, we prove that

$$w^\delta \in X_P$$

for every $\delta > 0$. In fact,

$$w^\delta = x - \delta t \geq 0$$

because $t \leq 0$ and $x \geq 0$. Furthermore:

$$
\begin{aligned}
Aw^\delta &= A_\mathsf{P}(x_\mathsf{P} - \delta t_\mathsf{P}) + A_{\{k\}}\delta + A_{\{1,\dots,n+m\}\setminus(\{k\}\cup \mathsf{P})}0 \\
&= A_\mathsf{P}x_\mathsf{P} + \delta(A_{\{k\}} - A_\mathsf{P}t_\mathsf{P}) \\
&= A_\mathsf{P}x_\mathsf{P} + \delta(A_{\{k\}} - A_\mathsf{P}(A_\mathsf{P})^{-1}a_{\bullet k}) \\
&= A_\mathsf{P}x_\mathsf{P} + \delta(A_{\{k\}} - a_{\bullet k}) \\
&= A_\mathsf{P}x_\mathsf{P} \\
&= b.
\end{aligned}
$$

We now show that, for each $z \in X_P$ there is $\delta > 0$ such that

$$cw^\delta < cz.$$

Consider two cases:

(1) $cz < c_\mathsf{P}x_\mathsf{P}$. Let

$$\delta > \frac{cz - c_\mathsf{P}x_\mathsf{P}}{c_k - y^\mathsf{T}a_{\bullet k}} > 0.$$

Then,

$$
\begin{aligned}
cw^\delta &= c_P x_P + \delta(c_k - y^T a_{\bullet k}) && \text{by (\dag)}\\
&< c_P x_P + \frac{cz - c_P x_P}{c_k - y^T a_{\bullet k}}(c_k - y^T a_{\bullet k})\\
&= cz.
\end{aligned}
$$

(2) $cz \geq c_P x_P$. Then, $cw^\delta < cz$ since $cw^\delta < cx$ for every $\delta \in \mathbb{R}^+$, as seen above. Therefore, $S_P = \emptyset$. \hfill QED

Proposition 8.4

Let $P = (A, b, c) \in CS(\mathcal{C})$ be a non-degenerate problem with $b > 0$. Assume that $x \in X_P$. Then, in Step (4c)(ii) of the Simplex Algorithm in Figure 8.1, $x - ot \in X_P$ and $c(x - ot) < cx$.

Proof:

Let j be such that

$$
o = \frac{x_j}{t_j}.
$$

Note that $t_j > 0$. Then,

$$
j \in P,
$$

because $t_k = -1$ and $t_{\{1,\ldots,n+m\}\setminus(P\cup\{k\})} = 0$. Thus,

$$
x_j > 0.
$$

Thus,

$$
o > 0.
$$

Note that

$$
x - ot \text{ is } w^\circ,
$$

where w^δ was defined in the proof of Proposition 8.3. Then,

$$
x - ot \in X_P
$$

and

$$
c(x - ot) < cx,
$$

as already established in the proof of Proposition 8.3. \hfill QED

Proposition 8.5

Let $P = (A, b, c) \in CS(\mathcal{C})$ be a non-degenerate problem with $b > 0$. Then, in Step 1 and in each execution of Step (4c)(ii) of the Simplex Algorithm in Figure 8.1, x is a basic admissible vector.

Proof:

The proof follows by induction on the number k of assignments on x during the execution of s.

(Base) $k = 1$. Then, x is basic by Proposition 8.1.

(Step) Assume by induction hypothesis that x is a basic vector in the $(k-1)$-th execution of Step (4c)(ii) of the Simplex Algorithm in Figure 8.1. We must show that

$$x - ot$$

is a basic vector. Observe that $x - ot \in X_P$ by Proposition 8.4. By Proposition 4.9, because P is non-degenerate, it suffices to establish

$$|\{j \in \{1, \ldots, n + m\} : x_j - ot_j > 0\}| = m.$$

Observe that $|P_x| = m$ since x is basic. On the other hand,

$$P_{x-ot} = \{j \in \{1, \ldots, n + m\} : w_j^o > 0\}$$

taking into account the proof of Proposition 8.4. Let j be such that

$$o = \frac{x_j}{t_j}.$$

We now prove that

$$P_{x-ot} = \{k\} \cup (P_x \setminus \{j\}).$$

Since

$$k \notin P_x$$

(see the proof of Proposition 8.3), then

$$w_k^o = o > 0$$

(see the proof of Proposition 8.4). Hence,

$$k \in P_{x-ot}.$$

On the other hand, since $j \in P_x$ (taking into account that o > 0), then

$$x_j - ot_j = x_j - \frac{x_j}{t_j} t_j = 0.$$

Thus,

$$j \notin P_{x-ot}.$$

Note that

$$x_\ell - ot_\ell = 0,$$

for every $\ell \in \{1, \ldots, n+m\} \setminus (P_{\mathsf{x}} \cup \{\mathsf{k}\})$. Therefore,

$$\ell \in (\{1, \ldots, n+m\} \setminus P_{\mathsf{x}-\mathsf{ot}})$$

for every $\ell \in \{1, \ldots, n+m\} \setminus (P_{\mathsf{x}} \cup \{\mathsf{k}\})$. Hence,

$$|\{1, \ldots, n+m\} \setminus P_{\mathsf{x}-\mathsf{ot}}| \geq n-1.$$

On the other hand,

$$j \notin \{1, \ldots, n+m\} \setminus (P_{\mathsf{x}} \cup \{\mathsf{k}\})$$

since $j \in P_{\mathsf{x}}$. Thus,

$$|\{1, \ldots, n+m\} \setminus P_{\mathsf{x}-\mathsf{ot}}| \geq n.$$

since $j \notin P_{\mathsf{x}-\mathsf{ot}}$. So,

$$|P_{\mathsf{x}-\mathsf{ot}}| \leq m.$$

Assume, by contradiction, that

$$|P_{\mathsf{x}-\mathsf{ot}}| < m = |P_{\mathsf{x}}|.$$

Since $\mathsf{x} - \mathsf{ot} \in X_P$, by Proposition 8.4, then

$$b = \sum_{\ell \in P_{\mathsf{x}-\mathsf{ot}}} a_{\bullet \ell}(\mathsf{x} - \mathsf{ot})_{\ell}.$$

Thus, b would be a linear combination of less than m columns of A contradicting the non-degeneracy assumption.   QED

Proposition 8.6

Let $P = (A, b, c) \in CS(\mathcal{C})$ be a non-degenerate problem with $b > 0$. Then, the execution of the Simplex Algorithm in Figure 8.1 on P terminates.

Proof:

Let x^i be the basic vector of the i-th iteration of the execution of $\mathfrak{s}$ on P. Observe that

$$|\{\mathsf{x}^i : i \leq j\}| = j$$

for every $j \in \mathbb{N}^+$, since

$$c\mathsf{x}^{i+1} < c\mathsf{x}^i,$$

by Proposition 8.4. Assume, by contradiction, that the execution of $\mathfrak{s}$ on P does not terminate. Then,

$$|\{\mathsf{x}^i : i \in \mathbb{N}^+\}| = |\mathbb{N}|.$$

Therefore, $\{x^i : i \in \mathbb{N}^+\}$ is not finite. Moreover, by Proposition 8.5,

$$\{x^i : i \in \mathbb{N}^+\} \subseteq B_P,$$

where B_P is the set of basic vectors of P. Therefore, B_P is infinite contradicting Proposition 4.5. QED

We are ready to prove the *Soundness of the Simplex Algorithm Theorem*.

Theorem 8.1

Let $P = (A, b, c) \in CS(\mathcal{C})$ be a non-degenerate problem with $b > 0$. Then, the execution of the Simplex Algorithm in Figure 8.1 on P terminates

- either with $\mathsf{b} = 0$ and then $S_P = \emptyset$;

- or with $\mathsf{b} = 1$ and then $\mathsf{s} \in S_P$.

Proof:

(1) Assume that s terminates with $\mathsf{b} = 0$. Then, $\mathsf{x} \in X_P$, $\mathsf{t} \leq 0$ and the guard of the Simplex Algorithm in Step 4 is false. Therefore, by Proposition 8.3, $S_P = \emptyset$.

(2) Assume that s terminates with $\mathsf{b} = 1$. Then, $\mathsf{x} \in X_P$ and the guard of Step 4 of the Simplex Algorithm is true. Thus, by Proposition 8.2, $\mathsf{x} \in S_P$ and so $\mathsf{s} \in S_P$. QED

We now prove the *Completeness of the Simplex Algorithm Theorem*.

Theorem 8.2

Let $P = (A, b, c) \in CS(\mathcal{C})$ be a non-degenerate problem with $b > 0$. Then,

- if $S_P = \emptyset$ then the execution of s on P terminates with $\mathsf{b} = 0$;

- if $S_P \neq \emptyset$ then the execution of s on P terminates with $\mathsf{b} = 1$ and $\mathsf{s} \in S_P$.

Proof:

Let Q be the dual of P (see Exercise 6.2).

(1) Assume that $S_P = \emptyset$. Note that $X_P \neq \emptyset$ since the initial x is an admissible vector, by Proposition 8.1. Using Exercise 6.5, we conclude that:

$$Y_Q = \emptyset.$$

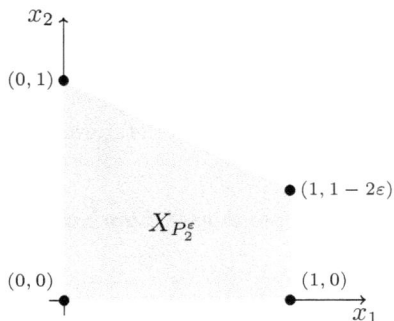

Figure 8.2: Set of admissible vectors of P_2^ε.

Observe that the guard of Step 4 tests whether or not y is an admissible vector of Q. Hence, the guard of Step 4 never holds since $Y_Q = \emptyset$. Hence, the execution of s on P never terminates with $(\mathsf{b}, \mathsf{s}) = (1, \mathsf{x})$. Since the execution of s on P terminates, by Proposition 8.6, then $\mathsf{b} = 0$.

(2) Assume that $S_P \neq \emptyset$. By Theorem 8.1, the execution of s on P does not terminate with $\mathsf{b} = 0$ but with $\mathsf{b} = 1$ and with $\mathsf{s} \in S_P$.   QED

The Simplex Algorithm is not efficient in general. Indeed, it is enough to consider a family of problems such that the execution of the simplex on the n-th problem of the family requires $2^n - 1$ iterations (along the lines of Proposition 7.10). We use a variant of the Klee-Minty example, see [44, 52], as the family of problems.

Definition 8.2
Let $0 < \varepsilon \leq \frac{1}{3}$ and $n \in \mathbb{N}^+$. The standard optimization problem P_n^ε is defined as follows:

$$
\begin{cases}
\displaystyle\min_x -\sum_{j=1}^n \varepsilon^{n-j} x_j \\[2mm]
x_1 + x_{n+1} = 1 \\[2mm]
\displaystyle 2\sum_{j=1}^{i-1} \varepsilon^{i-j} x_j + x_{n+i} = 1, \quad i = 2, \ldots, n \\[2mm]
x \geq 0.
\end{cases}
$$

For instance, the set of admissible vectors of the canonical optimization problem corresponding to P_2^ε is depicted in Figure 8.2 as well as the basic vectors. As shown in [52], P_n^ε has 2^n basic admissible vectors and the Simplex Algorithm when executed on P_n^ε requires $2^n - 1$ iterations using an appropriate choice of k. Therefore, the Simplex Algorithm visits all the basic admissible vectors.

8.3 Solved Problems and Exercises

Problem 8.1

Let P be the following optimization problem:

$$\begin{cases} \min_{x} \begin{bmatrix} -2 & -1 & 0 & 0 \end{bmatrix} x \\[2mm] \begin{bmatrix} -1 & 1 & 1 & 0 \\ 1 & -1 & 0 & 1 \end{bmatrix} x = \begin{bmatrix} 1 \\ 1 \end{bmatrix} \\[2mm] x \geq 0. \end{cases}$$

Show that $S_P = \emptyset$ using the Simplex Algorithm in Figure 8.1.

Solution:

We start by noting that $P \in CS(\mathcal{C})$, P is non-degenerate and $b > 0$. Hence, we can apply the Simplex Algorithm. Let Q be the dual of P. The initial basic admissible vector x is

$$\begin{bmatrix} 0 \\ 0 \\ 1 \\ 1 \end{bmatrix}.$$

Moreover, P $= \{3, 4\}$. The unique solution of

$$(A_P)^\mathsf{T} y = (c_P)^\mathsf{T}$$

is y $= (0, 0)$. Since

$$A^\mathsf{T} y \not\leq c^\mathsf{T}$$

for the first and the second components, then y $\notin Y_Q$. Take k $= 1$. Then,

$$t = \begin{bmatrix} -1 \\ 0 \\ -1 \\ 1 \end{bmatrix}.$$

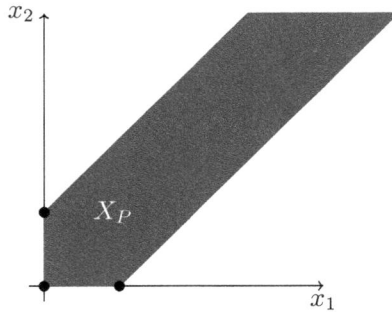

Figure 8.3: Set $X_{P'}$ where $CS(P') = P$ with P in Problem 8.1.

Since it is not the case that $\mathsf{t} \leq 0$, then $\mathsf{o} = 1$. Hence,

$$\mathsf{x} = \begin{bmatrix} 1 \\ 0 \\ 2 \\ 0 \end{bmatrix}.$$

Then, $\mathsf{P} = \{1, 3\}$. The unique solution of

$$(A_{\mathsf{P}})^{\mathsf{T}} y = (c_{\mathsf{P}})^{\mathsf{T}}$$

is $\mathsf{y} = (0, -2)$. Since

$$A^{\mathsf{T}} \mathsf{y} \nleq c^{\mathsf{T}}$$

for the first and the second components, then $\mathsf{y} \notin Y_Q$. Hence, take $\mathsf{k} = 2$. Then,

$$\mathsf{t} = \begin{bmatrix} -1 \\ -1 \\ 0 \\ 0 \end{bmatrix}.$$

Since it is the case that $\mathsf{t} \leq 0$, then $\mathsf{b} = 0$. So, by Theorem 8.1, $S_P = \emptyset$.   ◁

Problem 8.2

Let P be the following optimization problem:

$$\begin{cases} \min\limits_{x} -2x_1 - x_2 \\ x_1 + x_3 = 5 \\ 4x_1 + x_2 + x_4 = 25 \\ x \geq 0. \end{cases}$$

Using the Simplex Algorithm provide an optimizer of P.

Solution:
It is immediate that P is non-degenerate and $b > 0$. Hence, we can apply the Simplex Algorithm. Let Q be the dual of P. The initial basic admissible vector x is

$$(0, 0, 5, 25).$$

Then, $\mathsf{P} = \{3, 4\}$. The unique solution of

$$(A_\mathsf{P})^\mathsf{T} y = (c_\mathsf{P})^\mathsf{T}$$

is $\mathsf{y} = (0, 0)$. Since

$$A^\mathsf{T} \mathsf{y} \not\leq c^\mathsf{T}$$

for the first and the second components, then $\mathsf{y} \notin Y_Q$. Hence, take $\mathsf{k} = 1$. Then,

$$t = \begin{bmatrix} -1 \\ 0 \\ 1 \\ 4 \end{bmatrix}.$$

Moreover,

$$\mathsf{o} = \min \left\{ 5, \frac{25}{4} \right\} = 5.$$

Therefore,

$$\mathsf{x} = (5, 0, 0, 5).$$

Then, $\mathsf{P} = \{1, 4\}$. The solution of

$$(A_\mathsf{P})^\mathsf{T} y = (c_\mathsf{P})^\mathsf{T}$$

is $\mathsf{y} = (-2, 0)$. Since

$$A^\mathsf{T} \mathsf{y} \not\leq c^\mathsf{T}$$

for the second component, then $\mathsf{y} \notin Y_Q$. Hence, take $\mathsf{k} = 2$. Then,

$$t = \begin{bmatrix} 0 \\ -1 \\ 0 \\ 1 \end{bmatrix}.$$

Moreover,

$$\mathsf{o} = 5.$$

Figure 8.4: Basic vectors visited in Example 8.2.

Thus,
$$\mathsf{x} = (5, 5, 0, 0).$$

Hence, $\mathsf{P} = \{1, 2\}$. The solution of
$$(A_\mathsf{P})^\mathsf{T} y = (c_\mathsf{P})^\mathsf{T}$$

is $\mathsf{y} = (2, -1)$. Since
$$A^\mathsf{T} \mathsf{y} \not\leq c^\mathsf{T}$$

for the third component, then $\mathsf{y} \notin Y_Q$. Hence, take $\mathsf{k} = 3$. Then,
$$\mathsf{t} = \begin{bmatrix} 1 \\ -4 \\ -1 \\ 0 \end{bmatrix}.$$

Moreover,
$$\mathsf{o} = 5.$$

Thus,
$$\mathsf{x} = (0, 25, 5, 0).$$

Hence, $\mathsf{P} = \{2, 3\}$. The solution of
$$(A_\mathsf{P})^\mathsf{T} y = (c_\mathsf{P})^\mathsf{T}$$

is $\mathsf{y} = (0, -1)$. This vector is in Y_Q. Therefore, $\mathsf{b} = 1$ and $\mathsf{s} = (0, 25, 5, 0)$ is an optimizer, by Theorem 8.1. ◁

Exercise 8.1

Let P be the following optimization problem:

$$\begin{cases} \min_{x} \ 3x_1 + 3x_2 + 2x_3 \\ x_1 + 2x_2 + 3x_3 = 3 \\ 4x_1 + 5x_2 + 6x_3 = 9 \\ x \geq 0. \end{cases}$$

Using reductions and the Simplex Algorithm find an optimizer of P, if there is one.

Chapter 9

Integer Optimization

An *integer optimization problem* is a linear optimization problem where the components of the admissible vectors are natural numbers and the components of A, b and c are rational numbers (recall Definition 1.7). More precisely, an n-dimensional *integer standard optimization problem* is a linear optimization problem of the form:

$$\begin{cases} \min_{x} cx \\ Ax = b \\ x \in \mathbb{N}^n, \end{cases}$$

where the components of A, b and c are assumed to be rational numbers. Similarly, we can define an n-dimensional *integer canonical optimization problem* as

$$\begin{cases} \max_{x} cx \\ Ax \leq b \\ x \in \mathbb{N}^n, \end{cases}$$

where the components of A, b and c are assumed to be rational numbers.

The reader can get a picture of the history of integer optimization in [41].

Example 9.1 (Knapsack Problem)
The Knapsack Problem consists in maximizing the value of objects that can be put in a knapsack assuming that objects have a value and a weight and that the weight of all the objects that can be carried in the knapsack has an upper bound. Suppose that we have different kinds of objects, say $j = 1, \ldots, n$. The

problem can be modeled as follows:

$$\begin{cases} \max_{x} cx \\ ax \le b \\ x \in \mathbb{N}^n, \end{cases}$$

where

- x_j is the number of objects of kind j;

- a_j is the weight of each object of kind j;

- c_j is the value of each object of kind j;

- b is the upper bound on the weight;

and a_j and c_j are positive rational numbers for $j = 1, \ldots, n$ and b is a positive rational number.

9.1   Relaxed Problem

Some of the techniques for solving an integer optimization problem employ the use of a companion optimization problem.

Definition 9.1
Let

$$P = \begin{cases} \max_{x} cx \\ Ax \le b \\ x \in \mathbb{N}^n \end{cases}$$

be an integer canonical optimization problem. The *relaxed canonical optimization problem* of P, denoted by

$$R(P),$$

is

$$\begin{cases} \max_{x} cx \\ Ax \le b \\ x \ge 0. \end{cases}$$

Similarly for an integer standard optimization problem.

It is worthwhile to compare the sets of admissible vectors and optimizers of P and $R(P)$.

Proposition 9.1
Let P be an integer canonical optimization problem. Then,

1. $X_P \subseteq X_{R(P)}$;

2. if $s \in S_P$ and $s' \in S_{R(P)}$ then $cs \leq cs'$;

3. if $S_{R(P)} \cap \mathbb{N}^n \neq \emptyset$ then $S_P \subseteq S_{R(P)}$.

Proof:
(1) Let $x \in X_P$. Then, $x \geq 0$ since $x \in \mathbb{N}^n$, and $Ax \leq b$. Therefore, $x \in X_{R(P)}$.

(2) Assume that $s \in S_P$ and $s' \in S_{R(P)}$. Observe that, by (1),

$$\max_x \{cx : x \in X_{R(P)}\} \geq \max_x \{cx : x \in X_P\}.$$

Then,

$$cs' = \max_x \{cx : x \in X_{R(P)}\} \geq \max_x \{cx : x \in X_P\} = cs.$$

(3) Assume that $s \in S_{R(P)} \cap \mathbb{N}^n$. Hence, $s \in X_P$. Moreover, $s \in S_P$ by (2). Let $s' \in S_P$. Then, $cs = cs'$ and so $s' \in S_{R(P)}$. QED

Another way to compare the sets of optimizers is via the integrality gap.

Definition 9.2
Let P be an integer canonical optimization problem such that there is $s' \in S_{R(P)}$ with $cs' \neq 0$. The *integrality gap* between S_P and $S_{R(P)}$, denoted by

$$\mathrm{IG}_P,$$

is

$$\frac{cs}{cs'},$$

where $s \in S_P$ and $s' \in S_{R(P)}$. Similarly, for integer standard optimization problems.

Observe that

$$\mathrm{IG}_P \begin{cases} \leq 1 & \text{if } P \text{ is canonical} \\ \geq 1 & \text{if } P \text{ is standard.} \end{cases}$$

The integrality gap is 1 when an optimizer of the relaxed problem is also an optimizer of the integer problem. In Proposition 9.7, we will see that when the restriction matrix of P is totally unimodular and P satisfies some additional mild conditions then

$$IG_P = 1$$

meaning that we do not need to introduce new techniques for dealing with such integer optimization problems.

Example 9.2 (Knapsack Problem)
Recall the knapsack problem P introduced in Example 9.1. The difficulty of the knapsack problem has to do with the $x_1, \ldots, x_n \in \mathbb{N}$ constraint. The relaxed canonical optimization problem $R(P)$ is

$$\begin{cases} \max_x cx \\ ax \leq b \\ x \geq 0 \end{cases}$$

and can be solved using duality techniques. Observe that the dual problem $Q_{R(P)}$ is

$$\begin{cases} \min_y by \\ a^\mathsf{T} y \geq c^\mathsf{T} \\ y \geq 0. \end{cases}$$

Let $k = 1, \ldots, n$ be an index such that

$$\frac{c_k}{a_k} = \max \left(\frac{c_1}{a_1}, \ldots, \frac{c_n}{a_n} \right)$$

(observe that the weight a_j and the value c_j of each object of kind j are positive, see Example 9.1); that is, k is a kind of an object with the greatest ratio value/weight. Let x be such that

$$x_j = \begin{cases} 0 & \text{if } j \neq k \\ \dfrac{b}{a_k} & \text{otherwise} \end{cases}$$

(note that b is positive, see Example 9.1); that is, we only put in the knapsack objects of kind k. Moreover, let

$$y = \frac{c_k}{a_k}.$$

We show that $x \in S_{R(P)}$. Indeed:

- $x \in X_{R(P)}$;

- $y \in Y_{Q_{R(P)}}$;

- $by = \dfrac{bc_k}{a_k} = cx$.

Therefore, by Proposition 6.4, $x \in S_{R(P)}$ and $y \in R_{Q_{R(P)}}$.

9.2 Totally Unimodular Problems

In this section, we intend to show that when the matrix of an integer canonical optimization problem is totally unimodular then, under suitable conditions, the integrality gap of the problem is 1 and the set of maximizers is non-empty.

Definition 9.3

An integer square matrix is *unimodular* if its determinant is either 1 or -1. A matrix is *totally unimodular* if each of its nonsingular submatrices (see Definition 1.23) is unimodular.

Proposition 9.2

The entries of a totally unimodular matrix are either 0 or 1 or -1.

Proof:

Let A be a totally unimodular matrix and a_{ij} a coefficient of A. Assume that $a_{ij} \neq 0$. Then, $[a_{ij}]$ is a nonsingular matrix with determinant a_{ij}. Since A is totally modular by hypothesis, then either $a_{ij} = 1$ or $a_{ij} = -1$. Otherwise $a_{ij} = 0$. \hfill QED

Example 9.3

For instance, the matrix

$$\begin{bmatrix} -1 & 0 & 1 \\ 1 & 1 & 0 \end{bmatrix}$$

is totally unimodular. Indeed:

(1) all the nonsingular submatrices 1×1 have determinants in $\{-1, 1\}$ since it is the case that det $[1] = 1$ and det $[-1] = -1$.

(2) all the nonsingular submatrices 2×2 have determinants in $\{-1, 1\}$ since it is the case that the determinant of each of the submatrices

$$\begin{bmatrix} -1 & 0 \\ 1 & 1 \end{bmatrix} \quad \begin{bmatrix} 0 & 1 \\ 1 & 0 \end{bmatrix} \quad \begin{bmatrix} -1 & 1 \\ 1 & 0 \end{bmatrix}$$

is -1.

Proposition 9.3
The collection of all unimodular $n \times n$-matrices is a group.

Proof:
(1) Assume that A and B are unimodular $n \times n$-matrices. Hence, det A and det B are in $\{-1, 1\}$. Then, $A \times B$ is also unimodular since

$$\det (A \times B) = (\det A) \times (\det B),$$

by Proposition 4.20. Therefore, $\det (A \times B) \in \{-1, 1\}$.

(2) I is unimodular since $\det I = 1$.

(3) Assume that A is a unimodular $n \times n$-matrix. Hence, $\det A \in \{-1, 1\}$. Then, A is nonsingular since $\det A \neq 0$ (see Proposition 4.21). Then, A^{-1} is also unimodular since

$$\det A^{-1} = \frac{1}{\det A},$$

by Proposition 4.20. Therefore, $\det A^{-1} \in \{-1, 1\}$.   QED

Proposition 9.4
Let A be an $m \times m$-matrix with integer coefficients. Then,

A is unimodular iff $\det A \neq 0$ and for each $b \in \mathbb{Z}^m$ if $Ax = b$ then $x \in \mathbb{Z}^m$.

Proof:
($\rightarrow$) Observe that A is a nonsingular matrix. Since $x = A^{-1}b$, then, by Cramer's Rule (see Proposition 4.23),

$$x_j = \frac{\det [A]_b^j}{\det A}.$$

Observe that either $\det A = 1$ or $\det A = -1$ and $\det [A]_b^j \in \mathbb{Z}$. Therefore, $x_j \in \mathbb{Z}$ for every $j = 1, \ldots, m$.

($\leftarrow$) Let b be the vector e^j (see Definition 1.20) and $x = A^{-1}e^j$. Then, by hypothesis, $x \in \mathbb{Z}^m$. Therefore,

$$(A^{-1})_{\bullet j} = A^{-1}e^j \in \mathbb{Z}^m$$

for every $j = 1, \ldots, m$. Thus, $\det A^{-1} \in \mathbb{Z}$. But

$$\det A^{-1} = \frac{1}{\det A}$$

by Proposition 4.20. Therefore, either $\det A^{-1} = \det A = 1$ or $\det A^{-1} = \det A = -1$. Hence, A is unimodular. QED

The following result states that total unimodularity is preserved under specific operations.

Proposition 9.5
Let A be a totally unimodular $m \times n$-matrix with integer coefficients. Then, any matrix obtained from A by adding a row or a column of the form $(0, \ldots, 1, \ldots, 0)$ is again totally unimodular. The same happens if we multiply any row or column of A by -1.

Proof:
Denote the coefficients of A by a_{ij} for any $i = 1, \ldots, m$ and $j = 1, \ldots, n$. We only consider the case where the matrix A' is obtained from A by adding the line $(0, \ldots, 1, \ldots, 0)$ at the top. That is, A' is the matrix:

$$\begin{bmatrix} 0 & \cdots & 1 & \cdots & 0 \\ a_{11} & \cdots & a_{1j} & \cdots & a_{1n} \\ \cdots & & & & \\ a_{m1} & \cdots & a_{mj} & \cdots & a_{mn} \end{bmatrix}$$

Let B' be a nonsingular submatrix of A'. There are two cases:
(1) B' is a submatrix of A. Then, $\det B' \in \{-1, 1\}$ since A is totally unimodular.
(2) B' is not a submatrix of A. Observe that B' should include element $a'_{1j} (= 1)$ of matrix A' because otherwise B' is singular. Hence,

$$\det B' = \pm \det B,$$

where B is the nonsingular submatrix of A composed by all lines of B' which the exception of line 1 and all columns of B' with the exception of column j of A' (see Definition 4.8). Since A is totally unimodular,

$$\det B \in \{-1, 1\}.$$

Hence, $\det B' \in \{-1, 1\}$. Therefore, A' is also totally unimodular. QED

We now provide a sufficient condition for a matrix to be totally unimodular (firstly introduced in [36]).

Proposition 9.6
Let A be a matrix such that:

- the entries of A are either 0 or 1 or -1;

- there are no more than two non-zero entries in each column;

- the set of the indices of the rows can be partitioned in two subsets with the following properties:

    - if a column has two entries of the same sign then the indices of the rows of those entries belong to different subsets of the partition;

    - if a column has two entries of different signs then the indices of the rows of those entries belong to the same subset of the partition.

Then, A is totally unimodular.

Proof:
We must show that each nonsingular $k \times k$-submatrix of A is unimodular, for every $k \in \mathbb{N}$. The proof is by induction on k.

Base: $k = 0$. Observe that there is a unique 0×0-submatrix of A, the empty matrix, which has determinant 1 (see Definition 4.8).

Step: Let $k = \ell+1$. Assume, by the induction hypothesis, that each nonsingular $\ell \times \ell$-submatrix of A has determinant either equal to 1 or equal to -1. Let A' be a nonsingular $k \times k$-submatrix of A. Recall (see Remark 4.2) that

$$\det A' = \sum_{j=1}^{k}(-1)^{i+j}a'_{ij}M^{A'}_{i,j},$$

where $M^{A'}_{i,j}$ is the (i,j)-minor of A'. There are three cases to consider:

(1) A' has a column with only zero entries. In this case, $\det A' = 0$. Hence, A' is a singular matrix. So, this case is not possible.

(2) A' has a column j with a unique non-zero entry. So, $a'_{ij} \in \{-1,1\}$. Then,

$$\det A' = \sum_{i=1}^{k}(-1)^{i+j}a'_{ij}M^{A'}_{i,j} = (-1)^{i+j}a'_{ij}M^{A'}_{i,j}.$$

Since A' is nonsingular, then $M_{i,j}^{A'} \neq 0$ and the matrix $A'_{i,j}$ is nonsingular. Thus, by the induction hypothesis, $M_{i,j}^{A'} \in \{-1, 1\}$. Therefore,

$$\det \ A' \in \{-1, 1\}.$$

(3) All columns of A' have two non-zero entries. Let $\{K_1, K_2\}$ be the partition of $\{1, \ldots, k\}$ corresponding to the given partition of A. We now show that

$$(\dagger) \quad \sum_{i \in K_1} a'_{ij} = \sum_{i \in K_2} a'_{ij}$$

for every $j = 1, \ldots, k$. Assume that $a'_{i_1 j}$ and $a'_{i_2 j}$ are the two non-zero entries of column j. Then, there are two cases:

(a) $a'_{i_1 j}$ and $a'_{i_2 j}$ have the same sign. Assume without loss of generality, that $i_1 \in K_1$, $i_2 \in K_2$ and $a'_{i_1 j} = a'_{i_2 j} = 1$. Then,

$$\sum_{i \in K_1} a'_{ij} - \sum_{i \in K_2} a'_{ij} = a'_{i_1 j} - a'_{i_2 j} = 1 - 1 = 0.$$

(b) $a'_{i_1 j}$ and $a'_{i_2 j}$ have different signs. Assume without loss of generality, that $i_1, i_2 \in K_1$, $a'_{i_1 j} = -1$ and $a'_{i_2 j} = 1$. Then,

$$\sum_{i \in K_1} a'_{ij} - \sum_{i \in K_2} a'_{ij} = \sum_{i \in K_1} a'_{ij} = a'_{i_1 j} + a'_{i_2 j} = -1 + 1 = 0.$$

Hence, by $(\dagger)$,

$$\sum_{i \in K_1} a'_{i\bullet} - \sum_{i \in K_2} a'_{i\bullet} = 0$$

meaning that the set of rows of A' is linearly dependent. Thus, $\det \ A' = 0$. Hence, A' is a singular matrix. So, this case is not possible. QED

Example 9.4
The matrix

$$\begin{bmatrix} 1 & 0 & 0 \\ 0 & -1 & 1 \\ 1 & 1 & -1 \end{bmatrix}$$

is totally unimodular. Indeed, it satisfies the first two conditions of Proposition 9.6. The partition $\{\{1\}, \{2, 3\}\}$ of the indices of rows shows that the third condition of Proposition 9.6 is also satisfied. Therefore, Proposition 9.6 allow us to conclude that the matrix is totally unimodular.

Example 9.5 (Assignment Problem)

The *Assignment Problem* consists in assigning to each individual in a collection V_1 of individuals a task from a collection V_2 of tasks in such a way that no task is performed by more than one individual and no individual executes more than one task. Moreover, we want to maximize the global satisfaction of the individuals on the assignment.

We assume that initially, each individual in V_1 had expressed which tasks in V_2 he/she is willing to perform and the respective degree of satisfaction. This can be modeled by

- a bipartite graph $(V, E)^1$ induced by the partition $\{V_1, V_2\}$ of V, where an edge $e \in E$ from $v_1 \in V_1$ to $v_2 \in V_2$ states that individual v_1 is willing to perform task v_2;

- a satisfaction degree map $c : E \to \mathbb{N}$ such that $c(e)$ is the degree of satisfaction of individual $S(e)$ in performing task $T(e)$.

Each feasible assignment of individuals to tasks can be seen as a map

$$x : E \to \{0, 1\}$$

such that,

- if $x(e) = 1$ then $x(e') = 0$ for every $e' \in E$ such that $S(e) = S(e')$ expressing that to each individual at most one task is assigned;

- if $x(e) = 1$ then $x(e') = 0$ for every $e' \in E$ such that $T(e) = T(e')$ expressing that to each at task at most one individual is assigned.

So, the assignment problem consists in finding a feasible assignment of individuals to tasks that maximizes the degree of satisfaction of the whole set of individuals. That is, the assignment problem consists in finding a feasible assignment s such that for every other assignment x

$$\sum_{e \in E} c(e)x(e) \le \sum_{e \in E} c(e)s(e).$$

Therefore, representing the maps x and c as vectors with components x_e and c_e, respectively, for each $e \in E$, the assignment problem can be presented

---

[1]A *graph* is a pair (V, E), where V is a set whose elements are called *vertices* and $E \subseteq V \times V$ is a set whose elements are called *edges*. For each $e \in E$, $S(e)$ gives the first component of e and $T(e)$ the second component. A finite graph is a graph where V is a finite set. A *bipartite graph* (V, E) induced by the partition $\{V_1, V_2\}$ of V is a finite graph such that: (1) $E \subseteq V_1 \times V_2$; (2) for each $v \in V_1$ there is $e \in E$ such that $S(e) = v$; (3) for each $v \in V_2$ there is $e \in E$ such that $T(e) = v$.

as the following integer canonical optimization problem:

$$
\begin{cases}
\displaystyle \max_{x} \sum_{e \in E} c_e x_e \\[2ex]
\displaystyle \sum_{e \in \{e \in E : S(e) = v_1\}} x_e \leq 1, \text{ for each } v_1 \in V_1 \\[2ex]
\displaystyle \sum_{e \in \{e \in E : T(e) = v_2\}} x_e \leq 1, \text{ for each } v_2 \in V_2 \\[2ex]
x_e \in \{0, 1\} \text{ for each } e \in E.
\end{cases}
$$

Therefore, assuming that $V_1 = \{v_1, \ldots, v_k\}$, $V_2 = \{v_{k+1}, \ldots, v_m\}$ and $E = \{e_1, \ldots, e_n\}$, the restriction $m \times n$-matrix A has the following entries:

$$
a_{ij} = \begin{cases}
1 & \text{if } S(e_j) = v_i \text{ and } 1 \leq i \leq k \\
1 & \text{if } T(e_j) = v_i \text{ and } k+1 \leq i \leq n \\
0 & \text{otherwise,}
\end{cases}
$$

b is a vector with m ones and c is a row vector with n componentes. Observe that $a_{ij} = 1$, where i is such that $1 \leq i \leq k$ if and only if v_i is the source of edge e_j and $a_{ij} = 1$, where i is such that $k+1 \leq i \leq m$ if and only if v_i is the target of edge e_j. Hence, if $e_j = (v_{i_1}, v_{i_2})$ then

$$
a_{i_1 j} = a_{i_2 j} = 1
$$

and of course $v_{i_1} \in V_1$ and $v_{i_2} \in V_2$. Furthermore,

$$
a_{ij} = 0
$$

for every $i \in \{1, \ldots, m\} \setminus \{i_1, i_2\}$. Thus, in each column there are exactly two entries equal to 1. So A is totally unimododular by Proposition 9.6 by partitioning the set of indices of rows in those indexed by V_1 and those indexed by V_2.

Consider the following sets $V_1 = \{v_1, v_2, v_3\}$ and $V_2 = \{v_4, v_5, v_6\}$ of individuals and tasks, respectively and the set $E = \{e_1 : v_1 \to v_4, e_2 : v_1 \to v_5, e_3 : v_2 \to v_5, e_4 : v_2 \to v_6, e_5 : v_3 \to v_6, e_6 : v_3 \to v_4\}$ of preferences. Moreover, assume that the degree of satisfaction for each possible choice is 1. So, the corresponding assignment problem is the following 0-1 integer canonical optimization problem:

$$
\begin{cases}
\max_{x} \; [1 \; 1 \; 1 \; 1 \; 1 \; 1] \, [x_{e_1} \; x_{e_2} \; x_{e_3} \; x_{e_4} \; x_{e_5} \; x_{e_6}]^{\mathsf{T}} \\[4pt]
\begin{bmatrix}
1 & 1 & 0 & 0 & 0 & 0 \\
0 & 0 & 1 & 1 & 0 & 0 \\
0 & 0 & 0 & 0 & 1 & 1 \\
1 & 0 & 0 & 0 & 0 & 1 \\
0 & 1 & 1 & 0 & 0 & 0 \\
0 & 0 & 0 & 1 & 1 & 0
\end{bmatrix}
\begin{bmatrix}
x_{e_1} \\ x_{e_2} \\ x_{e_3} \\ x_{e_4} \\ x_{e_5} \\ x_{e_6}
\end{bmatrix}
\leq
\begin{bmatrix}
1 \\ 1 \\ 1 \\ 1 \\ 1 \\ 1
\end{bmatrix} \\[4pt]
x_{e_1}, x_{e_2}, x_{e_3}, x_{e_4}, x_{e_5}, x_{e_6} \in \{0, 1\}.
\end{cases}
$$

As noted above, the restriction matrix of the assignment problem is totally unimodular by choosing the partition of $V_1 \cup V_2$ as

$$\{\{1, 2, 3\}, \{4, 5, 6\}\}.$$

Observe that the assignment $x = (0, 1, 1, 0, 1, 0)$ is not an admissible vector since $x_{e_2} + x_{e_3} = 2 \nleq 1$. In this case, individuals v_1 and v_2 would both be assigned to task v_5. The assignments

$$(1, 0, 1, 0, 1, 0) \quad \text{and} \quad (0, 1, 0, 1, 0, 1)$$

are maximizers of the problem corresponding to the following assignments

$$v_1 \mapsto v_4, v_2 \mapsto v_5, v_3 \mapsto v_6 \quad \text{and} \quad v_1 \mapsto v_5, v_2 \mapsto v_6, v_3 \mapsto v_4,$$

respectively. Indeed, they are admissible. Moreover, since

$$\min(|V_1|, |V_2|) = 3$$

then each admissible vector should have at most three 1's.

The reader interested in assignment problems may consult [13, 28, 56, 16]. Also of interest to this subject are the works [17, 51, 42].

We now provide a sufficient condition for an integer canonical optimization problem to have optimizers.

Proposition 9.7
Let $P = (A, b, c)$ be an integer canonical optimization problem. Assume that:

- A is a totally unimodular $m \times n$-matrix;

- $b \in \mathbb{Z}^m$;

- $X_P \neq \emptyset$;

- $x \mapsto cx$ is bounded from above in $X_{R(P)}$.

Then, S_P is non-empty and $\mathrm{IG}_P = 1$.

Proof:

The restriction matrix of $CS(R(P))$ is

$$[A \ I]$$

which is also totally unimodular by Proposition 9.5. Since $x \mapsto cx$ is bounded from above in $X_{R(P)}$, then $x \mapsto [-c \ 0] \, x$ is bounded from below in $X_{CS(R(P))}$. Observe that $X_{R(P)} \neq \emptyset$ by Proposition 9.1. Hence, $X_{CS(R(P))} \neq \emptyset$. Therefore, by Theorem 4.1, we conclude that there is a basic admissible vector, say $s = (s_1, \ldots, s_{n+m})$, such that

$$s \in S_{CS(R(P))}.$$

Thus, there is $B \subseteq \{1, \ldots, m\}$ such that A_B is nonsingular and

$$A_B s_B = b.$$

Since $[A \ I]$ is totally unimodular, then A_B is unimodular. Hence, by Proposition 9.4, $s_B \in \mathbb{Z}^m$. Moreover, since $s \in X_{CS(R(P))}$, then $s_B \in \mathbb{N}^m$. Furthermore, $s \in \mathbb{N}^{n+m}$. Thus, $(s_1, \ldots, s_n) \in S_P$ and so $\mathrm{IG}_P = 1$.   QED

The result above tells us that integer optimization problems with totally unimodular matrices under some mild conditions can be analyzed using generic linear optimization techniques.

9.3 The Branch and Bound Technique

As an illustration of a technique for solving integer linear optimization problems, we present in this section, a version of the branch and bound technique proposed in [20] based on [45]. For a survey on techniques for addressing this problem, including the cutting-plane technique proposed in [35], the interest reader may consult [58].

An algorithm based on the branch and bound technique can be found in Figure 9.1. The algorithm invokes an auxiliary algorithm

---

input : (A, b, c)

1. $(\mathsf{b}, \mathsf{s}, \mathsf{o}) := \mathfrak{a}(A, b, c)$;
2. if $\mathsf{b} = 0$ then return 0;
3. $w := \{((A, b, c), (\mathsf{b}, \mathsf{s}, \mathsf{o}))\}$;
4. while $w \neq \{\}$ do
   (a) let $((A', b', c), (\mathsf{b}, \mathsf{s}, \mathsf{o}))$ be the first element of w;
   (b) remove the first element from w;
   (c) if each component of s is in $\mathbb{N}$ then return $(1, \mathsf{s})$;
   (d) let j be such that $\mathsf{s}_j \notin \mathbb{N}$;
   (e) let $A'^{\leq}$, $b'^{\leq}$ be obtained by adding e^j to A', $\lfloor \mathsf{s}_j \rfloor$ to b';
   (f) $(\mathsf{b}, \mathsf{s}, \mathsf{o}) := \mathfrak{a}(A'^{\leq}, b'^{\leq}, c)$;
   (g) if $\mathsf{b} = 1$ then add $((A'^{\leq}, b'^{\leq}, c), (\mathsf{b}, \mathsf{s}, \mathsf{o}))$ to w sorted by $(*)$;
   (h) let $A'^{\geq}$, $b'^{\geq}$ be obtained by adding $-e^j$ to A', $-\lceil \mathsf{s}_j \rceil$ to b';
   (i) $(\mathsf{b}, \mathsf{s}, \mathsf{o}) := \mathfrak{a}(A'^{\geq}, b'^{\geq}, c)$;
   (j) if $\mathsf{b} = 1$ then add $((A'^{\geq}, b'^{\geq}, c), (\mathsf{b}, \mathsf{s}, \mathsf{o}))$ to w sorted by $(*)$;
5. return 0.

Figure 9.1: Algorithm based on the branch and bound technique

that when executed over a canonical optimization problem (A, b, c) returns $(\mathsf{b}, \mathsf{s}, \mathsf{o})$ such that $\mathsf{b} = 0$ when $S_P = \emptyset$ and $\mathsf{b} = 1$, $\mathsf{s} \in S_P$ and $\mathsf{o} = c\mathsf{s}$ otherwise. The variable w is a list and not a set (elements can occur repeatedly and the position of an element in the list is relevant). So, it makes sense to refer to the first element of w as well as to remove an element in a particular position of the list. Moreover, in the algorithm, $(*)$ means in descending order of o and for the same value of o putting first the ones with only natural components.

Example 9.6

Herein, we illustrate the application of the branch and bound technique to the following integer linear optimization problem

$$P = \begin{cases} \max\limits_{x} -2x_1 + 5x_2 \\ -4x_1 + 7x_2 \leq 21 \\ 2x_1 + 5x_2 \leq 32 \\ x_1 \leq 6 \\ x \in \mathbb{N}^2. \end{cases}$$

The corresponding relaxed linear optimization problem $R(P)$ is

$$\begin{cases} \max\limits_{x} -2x_1 + 5x_2 \\ -4x_1 + 7x_2 \leq 21 \\ 2x_1 + 5x_2 \leq 32 \\ x_1 \leq 6 \\ x \geq 0 \end{cases}$$

which has the maximizer

$$\left(\frac{7}{2}, 5\right)$$

and objective value 18. Thus, the initial value of w is

$$\left\{ \left(R(P), \left(1, \left(\frac{7}{2}, 5\right), 18\right) \right) \right\}.$$

Since the first component of the maximizer is not a natural number, we consider the following subproblems:

$$P_1 = \begin{cases} \max\limits_{x} -2x_1 + 5x_2 \\ -4x_1 + 7x_2 \leq 21 \\ 2x_1 + 5x_2 \leq 32 \\ x_1 \leq 6 \\ x_1 \leq 3 \\ x \geq 0 \end{cases}$$

which has the maximizer

$$\left(3, \frac{33}{7}\right)$$

and objective value $\dfrac{123}{7}$ and

$$P_2 = \begin{cases} \max\limits_{x} -2x_1 + 5x_2 \\ -4x_1 + 7x_2 \leq 21 \\ 2x_1 + 5x_2 \leq 32 \\ x_1 \leq 6 \\ x_1 \geq 4 \\ x \geq 0. \end{cases}$$

which has the maximizer

$$\left(4, \frac{24}{5}\right)$$

and objective value 16. Thus, the value of w is

$$\left\{\left(P_1, \left(1, \left(3, \frac{33}{7}\right), \frac{123}{7}\right)\right), \left(P_2, \left(1, \left(4, \frac{24}{5}\right), 16\right)\right)\right\}.$$

In the next step consider the first element of w. Since the second component of the maximizer of P_1 is not a natural number, we can consider the following two subproblems of P_1:

$$P_{11} = \begin{cases} \max\limits_{x} -2x_1 + 5x_2 \\ -4x_1 + 7x_2 \le 21 \\ 2x_1 + 5x_2 \le 32 \\ x_1 \le 3 \\ x_2 \le 4 \\ x \ge 0 \end{cases}$$

which has the maximizer

$$\left(\frac{7}{4}, 4\right)$$

and objective value $\dfrac{33}{2}$ and

$$P_{12} = \begin{cases} \max\limits_{x} -2x_1 + 5x_2 \\ -4x_1 + 7x_2 \le 21 \\ 2x_1 + 5x_2 \le 32 \\ x_1 \le 3 \\ x_2 \ge 5 \\ x \ge 0 \end{cases}$$

which has $X_{P_{12}} = \emptyset$. Thus, the value of w is

$$\left\{\left(P_{11}, \left(1, \left(\frac{7}{4}, 4\right), \frac{33}{2}\right)\right), \left(P_2, \left(1, \left(4, \frac{24}{5}\right), 16\right)\right)\right\}.$$

In the next step consider the first element of w. Since the first component of the maximizer of P_{11} is not a natural number, we can consider the following two subproblems of P_{11}:

$$P_{111} = \begin{cases} \max\limits_{x} -2x_1 + 5x_2 \\ -4x_1 + 7x_2 \le 21 \\ 2x_1 + 5x_2 \le 32 \\ x_1 \le 3 \\ x_2 \le 4 \\ x_1 \le 1 \\ x \ge 0 \end{cases}$$

which has the maximizer

$$\left(1, \frac{25}{7}\right)$$

and objective value $\dfrac{111}{7}$ and

$$P_{112} = \begin{cases} \max\limits_{x} -2x_1 + 5x_2 \\ -4x_1 + 7x_2 \le 21 \\ 2x_1 + 5x_2 \le 32 \\ x_1 \le 3 \\ x_2 \le 4 \\ x_1 \ge 2 \\ x \ge 0 \end{cases}$$

which has the maximizer

$$(2, 4)$$

and objective value 16. Thus, the value of w is

$$\left\{ (P_{112}, (1, (2, 4), 16)), \left(P_2, \left(1, \left(4, \frac{24}{5}\right), 16\right) \right), \right.$$
$$\left. \left(P_{111}, \left(1, \left(1, \frac{25}{7}\right), \frac{111}{7}\right) \right) \right\}.$$

In the next step the execution of the algorithm ends returning $(1, (2, 4))$.

For details about the complexity of integer optimization algorithms the reader may consult [50].

9.4   Solved Problems and Exercises

Problem 9.1 (Vertex Coloring Problem, see [40])
Let $G = (V, E)$ be a finite graph. The map

$$g : V \rightarrow \{1, \ldots, k\}$$

is a *k-coloring* of G whenever $g(v_1) \neq g(v_2)$ for every $(v_1, v_2) \in E$. The graph G is *k-colorable* if there is a *k*-coloring of G. The objective is to find the smallest number of colors from $\{1, \ldots, k\}$ that provides a *k*-coloring of G. Model this problem as a linear optimization problem.

Solution:
Assume that $V = \{v_1, \ldots, v_\ell\}$. An admissible vector has the form

$$(x_1, \ldots, x_k, z_1^{v_1}, \ldots, z_k^{v_\ell})$$

meaning that

$$x_j = \begin{cases} 1 & \text{if color } j \text{ is used} \\ 0 & \text{otherwise} \end{cases} \qquad z_j^v = \begin{cases} 1 & \text{if vertex } v \text{ is colored with } j \\ 0 & \text{otherwise.} \end{cases}$$

Hence, the linear optimization problem

$$\begin{cases} \displaystyle\min_{x, z^{v_1}, \ldots, z^{v_\ell}} \sum_{j=1}^{k} x_j & \\[2mm] \displaystyle\sum_{j=1}^{k} z_j^v = 1 & \text{for each } v \in V \\[2mm] z_j^v + z_j^{v'} \leq 1 & \text{for each } (v, v') \in E \\[2mm] z_j^v \leq x_j & \text{for each } v \in V \text{ and } j = 1, \ldots, k \\[2mm] x_j \in \{0, 1\} & \text{for each } j = 1, \ldots, k \\[2mm] z_j^v \in \{0, 1\} & \text{for each } v \in V \text{ and } j = 1, \ldots, k \end{cases}$$

models the vertex coloring problem. Observe that when there are no optimizers, we conclude that the graph is not *k*-colorable. When there are optimizers, the value of the objective map for each optimizer is the *chromatic number* of G (the smallest number of colors needed to color the graph).   ◁

Exercise 9.1

Consider the problem

$$
I = \begin{cases}
\max_{x} \; 2x_1 + 3x_2 \\[4pt]
-\frac{2}{3}x_1 + x_2 \leq 1 \\[4pt]
\frac{8}{3}x_1 + x_2 \leq 16 \\[4pt]
x \in \mathbb{N}.
\end{cases}
$$

Use the branch and bound algorithm to conclude whether or not this problem has a maximizer.

Exercise 9.2 (Traveling Problem)

Consider a tourist that wants to visit a particular collection of cities minimizing the number of kilometers traveling, in such a way that each city is visited once and only once and return to the departure city. Model this problem as an integer linear optimization problem.

Bibliography

[1] M. Agrawal, N. Kayal, and N. Saxena. Primes is in P. *Annals of Mathematics*, 2:781–793, 2002.

[2] D. Alevras and M. W. Padberg. *Linear Optimization and Extensions.* Springer, 2001.

[3] T. M. Apostol. *Calculus. Vol. I: One-Variable Calculus, with an Introduction to Linear Algebra.* Blaisdell, second edition, 1967.

[4] T. M. Apostol. *Calculus. Vol. II: Multi-Variable Calculus and Linear Algebra, with Applications to Differential Equations and Probability.* Wiley, second edition, 1991.

[5] S. Arora and B. Barak. *Computational Complexity.* Cambridge University Press, 2009.

[6] M. Audin. *Geometry.* Springer, 2003.

[7] S. Axler. *Linear Algebra Done Right.* Springer, second edition, 1997.

[8] M. K. Bennett. *Affine and Projective Geometry.* Wiley, 1995.

[9] M. Berger. *Geometry I.* Springer, 2009.

[10] D. P. Bertsekas. *Network Optimization: Continuous and Discrete Models.* Athena Scientific, 1998.

[11] D. P. Bertsekas. *Convex Analysis and Optimization.* Athena Scientific, 2003.

[12] D. Bertsimas and J. Tsitsiklis. *Introduction to Linear Optimization.* Athena Scientific, 1997.

[13] G. Birkhoff. Three observations on linear algebra. *Univ. Nac. Tucumán. Revista A.*, 5:147–151, 1946.

[14] T. S. Blyth and E. F. Robertson. *Basic Linear Algebra*. Springer, 2002.

[15] W. C. Brown. *Matrices and Vector Spaces*. Marcel Dekker, 1991.

[16] R. Burkard, M. Dell'Amico, and S. Martello. *Assignment Problems*. Society for Industrial and Applied Mathematics, 2009.

[17] R. E. Burkard and E. Çela. Linear assignment problems and extensions. In *Handbook of Combinatorial Optimization, Supplement Vol. A*, pages 75–149. Kluwer, 1999.

[18] S. Cook. The importance of the P versus NP question. *Journal of the ACM*, 50(1):27–29, 2003.

[19] T. H. Cormen, C. E. Leiserson, R. L. Rivest, and C. Stein. *Introduction to Algorithms*. MIT Press, third edition, 2009.

[20] R. J. Dakin. A tree-search algorithm for mixed integer programming problems. *The Computer Journal*, 8:250–255, 1965.

[21] G. B. Dantzig. A theorem on linear inequalities. Technical report, Pentagon, 1948.

[22] G. B. Dantzig. Maximization of a linear function of variables subject to linear inequalities. In *Activity Analysis of Production and Allocation*, pages 339–347. Wiley, 1951.

[23] G. B. Dantzig. Reminiscences about the origins of linear programming. *Operations Research Letters*, 1(2):43–48, 1982.

[24] G. B. Dantzig. Origins of the simplex method. In *A History of Scientific Computing*, pages 141–151. ACM, 1990.

[25] G. B. Dantzig. *Linear Programming and Extensions*. Princeton University Press, reprint of the 1968 corrected edition, 1998.

[26] G. B. Dantzig. Linear programming. *Operations Research*, 50(1):42–47, 2002.

[27] G. B. Dantzig and A. Orden. Notes on linear programming: Duality theorems. Technical Report RM-1265, AD 114135, US Air Force Project Rand, 1953.

[28] U. Derigs, O. Goecke, and R. Schrader. Monge sequences and a simple assignment algorithm. *Discrete Applied Mathematics*, 15(2-3):241–248, 1986.

[29] I. Ekeland and R. Témam. *Convex Analysis and Variational Problems*. Society for Industrial and Applied Mathematics, 1999.

[30] J. Farkas. Theorie der einfachen Ungleichungen. *Journal für die Reine und Angewandte Mathematik*, 124:1–27, 1902.

[31] W. Fleming. *Functions of Several Variables*. Springer, second edition, 1977.

[32] J. N. Franklin. *Methods of Mathematical Economics*. Society for Industrial and Applied Mathematics, 2002.

[33] J. Gallier. *Geometric Methods and Applications: For Computer Science and Engineering*. Springer, second edition, 2013.

[34] M. R. Garey and D. S. Johnson. *Computers and Intractability*. W. H. Freeman and Co., 1979.

[35] R. E. Gomory. An algorithm for integer solutions to linear programs. In *Recent Advances in Mathematical Programming*, pages 269–302. McGraw-Hill, 1963.

[36] I. Heller and C. B. Tompkins. An extension of a theorem of Dantzig's. In *Linear Inequalities and Related Systems*, volume 38 of *Annals of Mathematics Studies*, pages 247–254. Princeton University Press, 1956.

[37] D. Hilbert and W. Ackermann. *Grundzüge der theoretischen Logik*. Springer, 1959.

[38] F. E. Hohn. *Elementary Matrix Algebra*. Dover Publications, 2002.

[39] K. Ireland and M. Rosen. *A Classical Introduction to Modern Number Theory*. Springer, 1990.

[40] T. R. Jensen and B. Toft. *Graph Coloring Problems*. Wiley, 1995.

[41] Jünger, T. M., Liebling, D. Naddef, G. Nemhauser, W. Pulleyblank, G. Reinelt, G. Rinaldi, and L. Wolsey, editors. *50 Years of Integer Programming 1958–2008*. Springer, 2010.

[42] S. N. Kabadi and A. P. Punnen. A strongly polynomial simplex method for the linear fractional assignment problem. *Operations Research Letters*, 36(4):402–407, 2008.

[43] Y. Katznelson and Y. R. Katznelson. *A (Terse) Introduction to Linear Algebra*, volume 44. American Mathematical Society, 2008.

[44] V. Klee and G. J. Minty. How good is the simplex algorithm? In *Inequalities, III*, pages 159–175. Academic Press, 1972.

[45] A. H. Land and A. G. Doig. An automatic method of solving discrete programming problems. *Econometrica*, 28:497–520, 1960.

[46] S. Lang. *Linear Algebra*. Springer, third edition, 1989.

[47] S. Lang. *Algebra*. Springer, third edition, 2002.

[48] J. R. Munkres. *Topology*. Prentice-Hall, second edition, 2000.

[49] M. W. Padberg. *Linear Optimization and Extensions*. Springer, 1999.

[50] C. H. Papadimitriou. On the complexity of integer programming. *Journal of the Association for Computing Machinery*, 28(4):765–768, 1981.

[51] C. Papamanthou, K. Paparrizos, N. Samaras, and K. Stergiou. Worst case examples of an exterior point algorithm for the assignment problem. *Discrete Optimization*, 5(3):605–614, 2008.

[52] K. Paparrizos, N. Samaras, and D. Zissopoulos. Linear programming: Klee–Minty examples. In C. A. Floudas and P. M. Pardalos, editors, *Encyclopedia of Optimization*, pages 1891–1897. Springer, 2009.

[53] P. Pedregal. *Introduction to Optimization*. Springer, 2003.

[54] M. Reid and B. Szendrői. *Geometry and Topology*. Cambridge University Press, 2005.

[55] R. T. Rockafellar. *Convex Analysis*. Princeton University Press, 1970.

[56] A. E. Roth, U. G. Rothblum, and J. H. Vande Vate. Stable matchings, optimal assignments, and linear programming. *Mathematics of Operations Research*, 18(4):803–828, 1993.

[57] W. Rudin. *Functional Analysis*. McGraw-Hill, 1991.

[58] A. Schrijver. *Theory of Linear and Integer Programming*. Wiley, 2000.

[59] I. R. Shafarevich and A. O. Remizov. *Linear Algebra and Geometry*. Springer, 2013.

[60] T. Terlaky and S. Z. Zhang. Pivot rules for linear programming: A survey on recent theoretical developments. *Annals of Operations Research*, 46/47(1-4):203–233, 1993.

[61] J. von Neumann. On a maximization problem. Technical report, Institute for Advanced Studies, Princeton, 1947.

[62] J. von Neumann and O. Morgenstern. *Theory of Games and Economic Behavior*. Princeton University Press, second edition, 1947.

[63] S. J. Wright. *Primal-Dual Interior-Point Methods*. Society for Industrial and Applied Mathematics, 1997.

List of symbols

Subject index

www.ingramcontent.com/pod-product-compliance
Lightning Source LLC
Chambersburg PA
CBHW060349200326

41519CB00011BA/2081